MODERN DICTIONARY

CYTO-AND HISTOCHEMISTRY

C. K. Shah
Retd. Professor of Botany
Botany Department
University School of Sciences
Gujarat University
Ahmedabad-380009

Discovery Publishing House
New Delhi-110002

First Published 1991

ISBN 81-7141-151-7

Published by :
Discovery Publishing House
4594/9, Darya Ganj
New Delhi 110 002

Printed in India

Laser Typesetting by : **B.H. Enterprises, East of Kailash-110 065.**
Printed at : **Efficient Offset Printers, New Delhi–110 028.**

DEDICATED TO THE PIOUS MEMORY OF
PROFESSOR PANCHANAM MAHESHWARI FRS
WHO ALWAYS ENCOURAGED A HIGH
STANDARD OF TECHNICAL EXCELLENCE
AND SKILL.

PREFACE

The cell and tissue chemistry have evolved in diverse directions. It now consists of a vast amount of literature, battalion of biochemicals, procedures and measuring techniques. It examines from fixatives, adhesives to specific dye-metabolite complexes to their stoichiometry. Old schedules change giving place to new ones and vice versa. Foundations of histology have opened the gateway for cyto - and histochemistry which can be led or adapted to the special requirements of cytophotometry. The same metabolite shows different reactions with various chemicals while the same dye responds differently to various metabolites.

The approach is multidisciplinary viz. physical and chemical, techniques, cell, tissue, organ and organelle performance and its quantitative measurements. It also includes various instruments used for quantifying the reaction product and contains the explanation of related terms.

In compiling my materials, I must acknowledge the profuse guidelines which I received from the personal letters of Professors Drs. Arrigoni, O. (Bari), Bohdan Slavik (Praha), Briarty, L.G. (Nottingham), O'Brien, T. P. (Clayton), Elizabeth G. Cutter (Davis), Gifford, E.M. Jr. (Davis), Guignard, J.L. (Paris), Joseph Arditii (Irvine), Kavaljion, L.J. (Sacramento), Lin Chang-gan (Yunnan), Lincoln Richard, G. (Long Beach), Lance, S. Evans (New York), Marian Michniewicz (Torun), Nagl, W. (Kaiserslautern), Sharma, A.K. (Calcutta), Smith, A.R.W. (London), and Yeoman, M.M. (Edinburgh).

To all these wizards, I wish to tender my most grateful thanks for their willingness with which they co-operated wit me in a picosecond.

No words are enough to express my gratitude to Dr. M. Katyal for his kindness and blessings, Miss Kollappan Saroja gave me full co-operation in the preparation of titles, searching the literature and library work.

I am grateful to the University Grants commission for financing my major research project on histo-chemistry. I cannot afford to forget the painstaking efforts of Shri I.R.Wasan-the publishers.

I am eagerly looking forward to hear of any errors or misrepresentations that have ushered in my interpretation.

(C.K. Shah)

Absorbance

It measures the amount of monochromatic light absorbed by the coloured end product. It is useful for the quantitative analysis of coloured compound-the histochemical preparation. It is expressed as logarithms of the reciprocal of transmitted light after absorption. It is in fact the optical density (O.D.) or extinction (E) i.e. log I_o/I_s = O.D. = E=Absorbance.The row data from the microdensitometer or cytophotometer are the absorbances. Its value is related to field size, section thickness and to the intensity of the chromogenic reaction. The high UV absorbance quantifies nucleic acids.

Acetic acid

It is an open chain carboxylic acid. It does not coagulate proteins but extract basic substances such as protein and amino acids. DNA is precipitated from its solution by acetic acid. It neither fixes nor removes carbohydrates. Its rapid penetration causes marked swelling and hardening of tissue. Fixation needs frequent wahes in alcohol and then store material in 70% alcohol. It is an ingredient of nearly 30 kinds of fixatives as it produces an acid-fixation image due to a decrease in isoelectric point of proteins. It is never used alone as a fixative.

Acetic anhydride-pyridine

Sites of protein bound carboxyl groups stain red to purple. The slides need immersion in hydroxynaphthoic acid hydrazide prior to acetic anhydride-pyridine treatment. The dye is fast blue B salt.

Acetone

For enzyme analysis cold acetone is a popular fixing fluid. Dry tissue which is obtained by its evaporation is subsequently stored in a vacuum cryostat. It is an excellent dehydrant and a substitute for ethyl alcohol, to prepare dehydrating grades. Its anhydrous state is used only for final stages to remove the last drops of water. The procedure with acetone is exactly the same as with ethyl alcohol.

Acetyl choline

It is the acetyl ester of choline. It is secreted at the ends of nerve fibres at the arrival of nerve impulses. It has been localised in plant cells also. It seems to take part in the mechanism of conduction.

Acetyl choline esterase (ACH)

It has been long studied by the methods of enzyme histochemistry and identified as a component of surface membrane preparations from the synapse.

Achiotin

It is an extract of achiote seeds (*Bixa orellana* L). Neutral fats, cholesterol esters and certain lipids stain an intense yellow which is selective and characteristic for these cell constituents. Colour variations or shades depend upon different chemical structures of cell storage inclusions.

Acid haematein

Phospholipid sites appear blue black or perfectly black. The tissue sections require mordanting with dichromate-calcium reagent. Subsequently the insoluble compound formed combines with haematoxylin to produce a blue black colour.

Acid phosphatase

Lead nitrate is used in the incubating medium as lead phosphate is insoluble at pH 5.0. The precipitates are converted by means of yellow ammonium sulphide into brown sulphide of lead. The procedure has been of immense use in tissues following irradiation and endosperm and embryo development in coconut, and wheat caryopses. Sites of acid phosphatase activity appear brownish black of lead sulfide.

Acid dyes

These are an anionic dyes. They combine with a metal and form salts. They stain substances with a basic functional group. Therefore they stain cytoplasm e.g. acid fuschin, methyl blue, eosin, orange G, congo red, aniline blue. Such dyes combine with the nucleoproteins at a low pH.

Acidic nuclear Proteins

Proteins stained with FCF fast green at 5.0 pH are the acidic nuclear proteins-chromosomin of Stedman and Stedman (1950).They are blocked by methylation which has no influence on staining upto pH 8.0. Only histones respons to fast green at pH 8.00. Acetylated and dilute acid treated slides do not show staining at pH 8.00. Thus these two treatments namely acetylation and dilute acid treatment serve as strong controls for basic proteins. The reaction is not completely negative at pH 5.00 after dilute alkali treatment. Due to their residual nature all acidic proteins are not alkali labile. Thus fast green at 5.00 pH specifically binds with acidic nuclear proteins. They are linked wit gene regulation.

Acridine orange

It is a vital stain and imparts fluorescent emissions of different colours to DNA and RNA. Under optimum conditions, cell walls fluoresce orange or green, cut in bright yellow orange, pollen exine bright orange, cytoplasm green, DNA bright yellow and RNA flame red or pink under a fluorescent microscope.

Acriflavine

It has a specific binding to the DNA as ascertained by DNase and RNase treatment of the living cells. Parenchyma or squamous cells showed and estimated dye to DNA ratio as 1:4. The whole cells population method for determining average values can be followed.

Acrolein

It is a reagent which shows a remarkably good fixation of cellular details. A 10% acrolein solutiom destroys 75% of the enzyme activity. Sections showed red-to-magenta colour for the distribution of proteins by incubation in 5% acrolein followed by Schiff's reagent staining.

Adenyl cyclase

Deposits of fine electron opaque brown black points indicate the enzyme activity. It has to be entirely missing from the controls. The test is performed by 0.5 M 5-adenylylimido diphosphate (AMP) -PNP as a substrate containing theophylline, magnesium sulphate and lead nitrate. The material is maize, bean roots and embryos.

Alcian blue

It contains copper phthalocyanine on each of the benzenoid rings. It is a blue pigment of high stability. Positively stained tissue areas confirm the presence of acidic carbohydrates. But when combined with chromotrope 2 R areas change the colour from red to blue.

Alcian green

It is extremely useful to use as a counterstain as it gives pleasing green back ground. It is rapid, simple and gives better contrast. The counterstaining with metanil yellow or picrofuchsin is not specific.

Aldehyde radicals

Structures containing aldehyde radicals stain bluish violet following chemical liberation of radicals by 1% *p*-phenylenediamine transferred to a 1% aqueous gold chloride solution. The method is highly selective and more specific than the leucofuchsin technics.

Alizarin red S

It is extracted from the roots of madder (*Rubia* sp) but it is presently a synthetic product. It is also an anionic dye and imparts non-specific pink colour to the sections. It forms insoluble coloured chelates with protoplasmic calcium. The gemma cells of ferns like *Vittaria* and *Onoclea* accumulate calcium.

Alkaline phosphatase

Phosphomonoesterase I

The enzyme splits off the phosphate from a suitable substrate. The liberated phosphate with calcium ions form insoluble calcium phosphate as precipitates to cobalt phosphate and finally to black cobalt sulphide. The test is valid and quantitative and is found primarily with the nuclei.

Allexan's stain

When the staining solution is acidified with a drop of glacial acetic acid, pollen wall of the exine becomes bright green and the cytoplasm red. Further aborted pollen grains are greener light green, non-abortive or fertile viable pollens become bright red. It was devised by Alexander, M.P., by combining malachite green wit orange G.

Alloxan

Like ninhydrin, alloxan is a good substitute for the production of an aldehyde. It visualises all the proteins in a cell. It is the protein bound amino acids that are visible/stained in a fixed material. A blue colour wit Schiff's reagent is the ultimate result.

Aluminium

Morin (natural yellow II) reacts with aluminium salts in neutral or acidic solution to form a colloidally dispersed fine salt which fluoresces green. Salt of Be, Zn, Ga and Se in extremely minute quantities form fluorescent compounds with morin. Woody tissues give better results.

Alpha cellulose

It is insoluble in 17.5% NaOH. The fibrous layer or bars of the endothecial anther wall mainly consist of α-cellulose and not lignin as revealed by absence of red staining of vertical bars by phloroglucinol - HCl, Manle's test, chlorine sulphate test and Schiff's reaction.

Amidol solution

Deposits of fine black silver indicate the sites of calcium carbonate or calcium phosphate. Sections are treated with $AgNO_3$ (0.5%), amidol (0.5%) and aqueous 1% sodium thiosulphate.

Ammoniacal silver carbonate

Basic proteins are stained light black. But affinity to stain decreases progressively with increasing pH. Control lies in the selective extraction at different pH of acidic solutions. Even 0.1 N sulphuric acid serves as an extraction medium for histones.

Ammoniacal silver nitrate (ASN)

Black and Ansley's (1965) method for detection of basic proteins is sharp and contrasting but the mechanism of staining is not known. It (product) varies in colour from yellowish to black.The yellow is associated with lysine rich while blackish with arginine rich histones. Nitrous acid commonly blocks amino groups of lysine obliterating yellow ammoniacal silver staining while the black stain is altered to much lesser extent. The nucleus in the pollen grain shows arginine rich while the generative cell contains lysine rich histones as stained by ASN.The cytoplasm of plant as well as animal cells does not at all take ASN stain.

Aniline blue

Decolourized aniline blue stain viewed using a fluorescent microscope detects callose yellow in pollen tubes. The dye is water soluble but insoluble in alcohol. In visible light callose stains brilliant blue. Cotton blue or resorcin blue (lacmoid) serve the same pathway.

Antipodal cells

These cells are situated at the chalazal end of the embryo sac. Their nuclei are rich in DNA. The endopolypoidy level of antipodal nuclei reaches even upto 1024 C as measured (revealed)- by the cytophotometer. It has a relatively high concentration of ascorbic acid.

Arbutin

It is a β -glucoside composed of hydroquinone and glucose. Its presence in Frensh pear (*Pyrus*) enables the plant to resist infection by fire-blight bacteria. Arbutin stains dark blue to purple in plant tissues of the family Pyrolaceae, Proteaceae, Saxifragaceae, Lilaceae, Ericaceae and Rosaceae. Other constituents of the cell such as chlorogenic acid, caffeic acid and quinic acid stain yellow-green or red or brown in colour. The reagent is composed of 0.2% *p*-phenylenediamine in 2 N NH_4OH.

Arginine

Agrinine-rich basic proteins are stained deep red by 2,4-dichloro-1-naphthol. In alkaline mounting medium, the colour fades slowly but in non-alkaline medium, the colour is less stable, quickly fading to orange and then finally to yellow.

Ascorbic acid (System I)

It is a strong reducing agent which is universally present in the cells in sharply varying concentrations in its detection. Fresh material is dipped in acidified silver nitrate reagent-stored in dark in the cold at 0-4°C and pH 2 to 2.5 for 24-72 hours-washed in 50% alcoholic ammonia. This test is quite specific and stoichiometric. Vitamin C has a significant derepressive action to mould morphological expressions. It forms complexes with $AgNO_3$ which gives tan colour to black coloured granules. This method settles all the problems of ascorbic localisation arising from soluble vitamin C. Controls run by copper sulphate, acetic acid (30%) and formaldehyde entirely prevent black precipitates.

Ascorbic acid (System II)

Blackened silver grains or brown red colloidal deposits inhabit the sites of ascorbic acid. The method fails to localise free ascorbic acid which diffuses rapidly from the plant organ sections or whole tissues. The test solution is 10% silver nitrate in 10% acetic acid in the dark.

Ascorbic acid (System III)

Black or tan coloured colloidal granules indicate the sites of ascorbic acid in plant organ sections. The reaction is performed by 10% silver nitrate in 3% acetic acid with crystal violet in absolute alcohol.

Astrablau

Cellulose cell walls and solely the pure cellulose turn intense blue while depending upon the lignification, other cell walls show sulphur yellow to brick red. The stronger lignified the cell walls are, the more intensive yellow product appears.

Astrazone red FG
Syn cyanine red

It stains cellulose walls blue and lignified cell walls red to pink.Thus plant cell walls develop differential colour.On treating with n-butanol, cellulose walls lose the blue product. Thus it becomes selective.

ATPase (System I)

Brown black precipitates of lead sulphide are deposited at the sites of enzyme activity but its total absence in the control material is marked. In plant tissues ATPase (adenosinetriphosphatase) activity is mainly associated with the plasma membrane and the cell wall. The substrate is made up of disodium salt of ATP, lead nitrate and magnesium sulphate.

ATPase (System II)

The enzyme in the section splits the phosphate from the ATP. Then a brownish-black precipitate of lead phosphate occurs. The frozen sections of a leaf or growing apices are incubated in a medium containing 2% lead acetate, 2.5% magnesium nitrate and 0.125% ATP. Homogenates and isolated fractions are easy to study.

Auromine O

It detects lipids when viewed using a fluorescent microscope. Unsaturated acidic waxes and its precursors fluoresce bright greenish yellow.

Autoradiography

When a radioactive emission (such as an electron) strikes a photographic film, it produces an exposure with higher number of particles, the brighter and sharper the exposure. The organism is fed or injected with a substance containing radioactive atoms. A tissue section of the organism is then exposed to a photographic film. Bright spots reveal the distribution of the radioactive substance. It is the only technique to study the everchanging aspects of cells. Thus it is possible to localise specific molecules. Number of grains are proportional to the rate and amount of synthesis of a specific molecule. This opens a versatile gateway of quantitative cytochemistry.

Azo-coupling by oxidative deamination

The protein bound aldehydes detected by azobenzenephenyl hydroxine sulphonic acid gave a purple complex of hydrazone condensation, 2,4-dinitrophenylhydrazine coupled with the stable diazotate of 5-nitro-*o*-anisidine produced bright red precipitates, among arylamines, H chicago, J and gamma acids produced red dye complex with stable diazotate of 5-nitro-*o*-anisidine.

Azo dye coupling for acid and alkaline phosphatases

The sites of alkaline phosphatase activity are brown with fast red TR and black with fast black B. The substrate is alphanaphthyl phosphate and contains a diazonium salt. Alpha naphthol is liberated by the enzyme which couples with the diazonium salt to form an insoluble product. Sites of acid phosphate became reddish brown while nuclei blue. The diazonium salt forms a coloured azo dye.

Azure A

Azure A plus eosin combination stain nuclei and cytoplasmic RNA blue. But a solution of 0.25% Azure A with only a drop of thionyl chloride can replace Schiff's reagent to stain DNA, blue-green.

Azure B

Lignified tissues stain a crystal clear blue green at pH 4. The other components of the cell wall remain without the stain. DNA will appear green blue, RNA will specify grey or dark blue or purple and sieve plates beautiful cherry red. Thus it works as polychromatic stain.

Basic dyes

The Chromophoric group consists of basic atoms - i.e. cations-forming the actively staining part. It combines with an acid. They stain nuclei and thus they are basophilic. Basic fuschin, crystal violet, methyl green, methylene blue, saffrain, azures and acridine red are basic dyes. They stain nucleus and chromosomes.

Basic fuchsine

An acidified alcoholic solution of basic fuchsin is a superb substitute for Schiff's reagent. The shade of nuclear staining is little yellowish with aldehyde groups of polysaccharides as well as DNA. Infrared (microscopy) studies point that for cellulose the reaction product is blue azonethine.

Battaglia's mixture

The mixture is extremely useful as a fixative for vegetable and animal cells. It contain absolute alcohol (95%), chloroform, glacial acetic acid, formaldehyde in (5:1:1:1) proportion respectively.It is well adapted for histochemical detection.

Benzidine test

It is a very sensitive test for pollen viability. Viable pollen takes a blue colour while non-viable or degenerating pollen does not become blue. Presently this test is not used very much as some pollen gave opposite results.

Benzpyrene-3,4 reaction

Lipids stain blue or whitish-blue but the stain fades very rapidly. Incubation for 2-3 days and then ringing the coverslip prevents immediate drying.

Benzyl benzoate

It is an excellent clearing agent after staining the entire mounts.It is not harmful to the most sensitive dyes. Sections stained by the Golgi method have been stored in it without any special precautionary methods and remained unchanged even after two years.

Beryllium B

The nuclei are stained clear red, and cytoplasm blue. Beryllium oxide (BeO) takes dark blue points. Other organic tissue components either remain unstained or become pink. To mask the interference of other metal ions, one g Na_2-EDTA is added to 100 mL of the solution. Specificity of the reaction is extremely high.

Beta (β) -glucuronidase

The enzyme activity is manifested by intense blue product while it is entirely absent from no substrate as a control. The incubation medum is naphthol AS-B 1-β-D-glucuronide in dimethylformamide. Test solution without the substrate is the control.

Bi-col method

Strong acidic compounds stain blue with colloidal iron or gold. the staining is confined to nucleic acids because of their phosphate groups and to sulphated polysaccharides. Weakly acidic compounds such as hyaluronic acid stain red to dark brown. This method is employed to differentiate weakly acidic polysaccharides from strongly acidic ones.

Biebrich scarlet

This dye is only slightly soluble in alcohol but is quickly soluble in water. Graded alkaline pH levels provide a selective method of localising and differentiating proteins rich in basic amino acids. The above sulphonated dis-azo dye stains tissue sections with an intense red colour and lacks internal salt formation. The principle is that the extinction values for staining by acid dyes serve to distinguish basic constituents of proteins. The dye has a strong affinity for nuclear chromatin in the presence of phototungstic acid.

Biological Stain Commission Inc

The chief aims are : to standardise the biological stains; to publish scientific data relating to the nature and use of biological stains; to issue statements of certification; to carry on investigations looking towards the perfecting of the supply and development of new uses; and to evolve specifications and test methods so that only stains of standard quality reach the scientific community.

Bismark's brown

Successful pollen grain preparations should show a green exine of sporopollenin, and a brown intine layer. When counter-stained with fast green the cytoplasm is green. Vegetative and generative nuclei show pleasing brown.

Bismuth

The granules of bismuth appear bright red and nuclei violet purple by staining the sections with methyl violet followed by the brucine reagent.

Bismuth subnitrate

Its alkaline solution reacts well with carbohydrate-rich ingredients of Golgi bodies and forms fine metal deposits. The Golgi vesicles in plant leaves show polarity with respect to the localization of the reactive deposits and thus the chemical works as an excellent tracer.

Biuret reaction

Red colour points to peptide linkages in lower protein forms while intense violet colour indicates peptide linkages in higher proteins. Ammonium salts if present interfere with the reaction.

Block staining

Simple and rapid method to stain nucleoli in vegetable cells e.g. roots of onion by 2% $AgNO_3$ revealed heterogeneity of the nucleolus structure. It delimits the black nucleus, light black circular outline of the nuclear envelope and the cell wall, the cytoplam light colloidal granular and unidentified infra nucleolar components.

BME-DPX

The β mercapto-ethanol (BME) used in conjuction with the mounting medium DPX effectively prevents fluorescent fading. Thus it preserves excellent morphology. It facilitates quantification of metabolites.

Boron

With 50% alcoholic alkaline tumeric acid boron in crystalline form is stained light red. Boron is crystallised by dilute HCl, concentrated potassium iodide and finally sections are treated with concentrated KCl.

Bromine

It works as a control reaction on the periodic acid. Schiff's (PAS) method for carbohydrates. A positive PAS reaction after the bromine treatment of sections points that the colour is due to carbohydrates and not to lipids.

Bromophenol blue

It stains proteins from yellow to green to blue. The colour change depends upon pH range from 3.0 to 4.6.

Bound lipids

Masked lipids reveal black stain by Sudan black B in shoot and root apices. These are covered by protein binding to the lipid. Unmasking is brought about by little igniting the stain on the section. The heat facilitates the entry of the stain to the lipid.

Butvar B-98

It is sticky hydrophilic, non-flammable granular powder. It is a superior section-ribbons adhesive with definite advantages over celloidin. The solution contans 250 mg of Butavar B-98 in 100 mL of chloroform.

C-banding chromosomes

In plants-*Lathyrus*, *Scilla* and *Allium*, the banding of the chromosomes was superior to those obtained by conventional Giemsa stain. The C-bands stained deep purple with pinacyanol chloride in sharp contrast to the light pink of interband regions.

Cajeput oil

It is superior mounting and clearing medium far better than clove oil. It is obtained form the plant - *Melaleuca leucadendron* L. a tree of northern Australia and the Malay Archipelago.

Calcium (System I)

With calcium sulphate, white or colourless monoclinic needles will be formed if calcium is present.

Calcium (System II)

By placing the tissue sections in silver nitrate, insoluble calcium phosphate or calcium carbonate if calcium is present, the silver replaces the calcium. It forms silver carbonate or silver phosphate. The detection of silver on photographic development is sharp and clear.

Calcofluor M2R

The cellulose is routinely detected by iodine sulphuric acid test. Another test lies in calcofluor M2R New or Photine HV. These are fluorochromes. Sites of cellulose fluoresce using ultra violet light excitation by fluorescence microscopy.

Calcofluor M2R New

It is a vital dye. Sites of cellulose fluoresce under a fluorescent microscope using ultraviolet excitation. It is used for cultured cells, fresh material etc. Only 0.1% aqueous solution for 30 seconds determines cellulose.

Callose

It is a polysaccharide chemically analysed as β 1-3 d-glucan. It occurs inside the sieve tubes, enveloping layer of developing microspore and megaspore mother cells, damaged cells wounds, culture of carrot callus etc. As soon as deposited, it can be reabsorbed. Aqueous solution of resorcine blue or a 0.05% solution of aniline blue in M/15 K_2HPO_4 is quite specific. The fluorescence denotes the callose presence. In *Ottelia* advancing pollen tubes show progressing callose plugs at intervals of time. By callose, the cells become isolated from neighbouring cells and impart autonomy. It brings the activity of the tissue to an end either permanently or seasonally.

Canada balsam

It is a gum used for preparing permanent preparations of microscopic objects. It dissolves in xylene and dries hard. Its refractive index is like that of proteins.

Callose by aniline blue

The stain emits characteristic radiation of low energy value when put under ultraviolet rays. The hydrated sections are treated with 0.1% w/v aniline blue in 0.1 M potassium orthophosphate. Pollen tubes show bright yellow fluorescence.

Carmine and Orcein

It is obtained by grinding the dried bodies of female insect known as Coccus. The structure of carmine is uncertain as it contains insect proteins, dye metal complex of carminic acid and aluminium. Like aceto-orcenin, acetocarmine is a superior stain for mitotic chromosome but its use in histochemical methods is problematic. Aceto-orce in stains chromosomes purplish to brownish red, but cytoplasm and nuclei acquire no colouration. Bloom has established its specificity.

Carnoy's fluid

It is a rapidly penetrative fixative which extracts lipids but coagulates proteins and nucleic acids. Peak fixation period is 6-8 h only. Pine nuclei showed DNA magenta red by Schiff's reagent. Absorption was linearly related to both section thickness and DNA concentration per nucleus. Amongst all Carnoy's fluid was clearly superior in that it facilitated vigorous staining and contributed least resistance to enzyme extraction.

Castor oil

It counter acts the weakening effect of 8-hydroxyquinoline to emphasize the constriction morphology of the chromosomes of red wood tree. Addition of a drop of castor oil improves the counting clearly.

Catalase

It breaks down the toxic and poisonous substance - hydrogen peroxide formed during cell metabolism to water and oxygen. In studying catalase activity, it behaves in an identical way to peroxidase in staining reaction. But only a short washing time is necessary for catalase and a long post-fixation wash for peroxidase. Guaiacol (*o*-hydroxy anisole) is used as an alternative to benzidine. Brown to black deposits indicate catalase activity.

Cationic chelate

It has been a bye-product prepared from gallocyanin-chrome alum. Structures containing nucleic acids take blue black colour. No differentiation is needed and the stain is self-limiting. Thus its staining mechanism and the specificity become more authentic.

Celestine blue B

It is a fine nuclear stain first of its kind. With a trace of a mordant like iron (as ferric alum) chromium or aluminium in solution the nuclei become studded with blue colloidal particles while the unchanged dye lightly stains the cytoplasm. This cytoplasmic staining is removed by alcohol during dehydration.

Cell fractionation

In the process, the cells become disrupted and nuclei stabilised while cytoplasm dissolves in citric acid solution. These fractionated nuclei are stained bright blue and nucleoli and cytoplasmic organelles bright pink when fresh suspensions or unfixed dried smears of tissue homogenates are treated with a mixture of methyl green, pyronin B, glycerine and aqueous phenol. Floating paraffin sections in warm water before final mounting even result in a loss of little nucleic acids or proteins.

Cell membrane surface

Mucoprotein glycolipid forms long filaments firmly attached to the lamellar surface membrane. At the level of light microscopy, the filaments are stained with the periodic acid Schiff's reagent. Acidic mucopolysaccharides are localised by the interaction with colloidal iron and thorium dioxide.

Cell vitality

Newly synthesized plant cell walls, *Fucus* eggs, algal cells, isolated protoplasts of tobacco leaf tissue and porous structures of primary cell walls take shining white colour with calcofluor white M2R. It absorbs short wavelengths of light and emits blue light in walls of living cells of staminal hairs of *Tradescantia* while dead cells show only bright stained nuclei.

Cellulase

The reducing sugars formed by the enzyme-substrate interaction reduce the alkalized (KOH) silver nitrate to form black silver oxide at the site of the enzyme action. The products of cellulolysis are reducing sugars which can reduce the salts of heavy metals. The substrate is the carboxymethyl cellulose.

Cellulose

It is a principal structural component of plant cell walls, in all green plants and in a few Fungi. It has a fibrous structure to which is due its use in textiles (artificial silk, linen, cotton etc). It stains with gentian violet or fast green but its nuclei stained red with saffranin give a sharp contrast. It is insoluble in all common reagents.

Central cell

This is the largest cell of the female gamettophyte. Numerous glyoxysomes containing enzymes for β-oxidation of fatty acids as well as lipids are localised in the cell.

Cerium for phosphatases

At alkaline pH calcium is used as a capturing agent and the precipitated calcium phosphate to cerium phosphate. At neutral and acid pH, cerium is used directly as a capturing agent. Cerium phosphate is visualised by H_2O_2-DAB method.

Chitin

Chitin is converted into chitosan by potassium hydroxide (25%). Now it is stained with potassium iodide (IKI) in 1% sulphuric acid to yield an elegant violet colour. Brown colour points to a negative response. The chitosan is soluble in 2% acetic acid but not cellulose.

Citinase

Only chitin is attacked by the enzyme fluorescent chitinase (FC) while other structural features of algae, fungi and lichenes remain unaltered. Conjugates of chitinase with fluoresce in isothiocyanate or lissamine rhodamine is a specific stain *in situ*.

Chloral hydrate formaldehyde fixative for enzymes

In chloral formal in fixative, β-glucuronidase is preserved for 3 weeks, acid phosphatase for 6 wks and esterase and lipase for 9 wks. Alkaline phosphatase remained active for only one day. Thus this fixative is extremely useful to fix and prolong the period of enzyme activity.

Chloramine - T Schiff's reaction

Total proteins stain purple. If the colour is obtained in either deamination or acetylation control, the reaction reveals some other compound but not proteins. It will not have α-amino and α-carboxyl groups in the constituent amino acids. The chlorine is supplied by sodium hypochlorite or chloramine-T.

Chlorantine fast green BLL

Cell wall appositions and sieve plates in the phloem complex tissues stain deep green with chlorantine fast green BLL. The location of stained material is exactly a photocopy of the sites of material stained with resorcinol blue and amiline blue, Oak phloem gave specific staining.

Chlorazol Black E

The chromosome and nucleoli are stained an intense black and stood out clearly against the light grey cytoplasm and dark grey nucleus. The mitotic figures show up very nicely and hence it is extremely useful for making chromosome counts in plants. It is sulphonated tris-azo, acidic dye. Methyl cellosolve is used as a solvent for the dye. It is a stain for tension wood and particularly for gelatinous fibres in sections. It works as a complementary stain for lignin and it becomes pink.

Chlorazol paper brown B

It is a chlorazol dye used as a selective stain. In the plant of *Clematis vitalba* L, the epidermis and cortex stain yellow, pericycle blood red, xylem salmon to blood red, primary phloem orange, cambium pale yellow, secondary phleoem crimson red, pith amber-coloured and sieve plates bright crimson.

Chloride

It is detected as dark black granules of AgCl by treatment with 2% $AgNO_3$ and 2% HNO_3. Thus the method is based on the insolubility of AgCl in acid solution.

Chlorine-Sulphite

All xylem elements, fibres and sclerenchyma stain bright red when sections are flooded with calcium hypochlorite followed by sodium sulphate. After one hour, the colour starts fading to brown.

Chlorus acid ($HClO_2$)

It converts aldehyde groups to carboxylic ones. This blocking reaction is irreversible and quantitative without any steric hindrance. It is extremely useful to stabilise and block tissue aldehydes.

Chrome-alum-galloxyanine

Dye gallocyanine is boiled wit chrome-alum to preform a dye-metal complex. The bound complex is surprisingly stable towards organic solvents, water and light. Thus it provides a useful tool in quantitative microphotometric studies of the nucleic acid contents of cells. DNA is bluish green to green and RNA is pink to red.

Chrome-alum-gelatin

It has been found as a far superior and excellent adhesive to results given by Haupt and Mayer's adhesives. The stubbing solution is highly resistant to acids and bases and remains unstained throughout the staining schedule.

Chromic acid

It finds a poor place as fixative. In CRAF it is mixed with acetic acid to counteract shrinkage and hardening, chemical reactions with tissue constituents remain in oblivion. McCrae (1967) does not find any useful purpose with it. The affinity of proteins for basic dyes is extremely low. It effectively blocks the -COOH groups of the acidic amino acids and nuclear acidic proteins are dissolved out of the tissue.

Chromium trioxide

One percent Cr_2O_3 with formalin is a mighty oxidising agent, penetrating slowly, shrinking tissues with little hardening chromosome morphology is quite well preserved, and only DNA is precipitated in an insoluble form.

Chromosomal histones

Glutaraldehyde fixed material treated with ammoniacal silver carbonate formaldehyde staining imparts bluish black colour to the sites of high histone concentrations in mitotic chromosomes of cell preparations or monolayer cultures. Cells treated with cold. 0.25 N HCl before fixation are not at all stained. But chromatin does not reduce silver by itself.

Chromyl chloride

On prolonged exposure to chromyl chloride at room temperature a typical Feulgen nucleal reaction is obtained in tissue sections. Thus it is used as a substitute supporting stain.

Chrysoidin yellow

It is an excellent dye which contrasts well with basic fuchsin. It stains chromatin shining yellow. It is made up and used in the same manner as common Feulgen reagent. Squash preparations are extremely useful for comparative measurements of nuclei and chromatin stained with chrysoidin yellow, basic fuchsin, toluidine O and Azure A.

***Citurs* vesicles**

The fruit vesicles has a peripherally located lining of suberin which is stained red effectively by Sudan III and Sudan IV. These two dyes are specifically employed to stain red rubber in rubber plant tissues.

Clear seal

It is an adhesive of silicon rubber. It prevents detachment of sections. It is not opaque and withstands temperatures upto 260° C, to cure without shrinkage and is resistant to acid and alkali.

Coosmassie blue

Place the sections in 0.25% w/v Coomassie brilliant blue R 250 in 7% aqueous acetic acid. Remove the excess stain in 5% V/V acetic acid. Total proteins stain violet in colour. Coomassie brilliant blue R 250 stains proteins lovely blue.

Concanavalin A

It is monosaccharide binding protein. Sites of mucosubstances containing α-D mannosyl or α-D-glucosyl residues turn dark brown. Controls should be completely unstained, and exclude the possibility of non-specific staining of substances other than those sugar residues to which the lectin is bound specifically.

Congo red

It is an indicator stain when pH is less than 3.0. It stains amyloid bright pink to red but nuclei blue. It imparts a brilliant red colour to methyl cellulose.

Content per cell and concentration per unit area

The total mass M of absorbing substance is obtained by multiplying the concentration by the the volume i.e. M= Cal where a is the area in microspectrophotometry, or the area of the microscopic field, it follows that M/A=a/k or M=Aa/k. Thus M is proportional to the product of absorbance and area; neither the concentrational nor the thickness appear in this formula.

If Lambert-Beer's law is applicable, the result will be related to the concentration of the chromophore in different tissue samples. To obtain an estimate of relative total amount in a particular tissue, the figures for relative concentration are multiplied by tissue volume. Relative amounts per cell can be obtained from this value by dividing by cell number. In a simple way the extinction values are then multiplied by the corresponding cell area to obtain the total content of the metabolite per cell. In the same way the extinction values are divided by the cell area to obtain the concentration per unit area of the cell.

Control reactions

It is extremely important to prove and pursue that the tissue analysed chemically is almost the same as the tissue studied morphologically. It is necessary to run proper control slides at every step of the reaction. It embarks on the specificity of histochemical reactions and ultimately supports its stoichiometry

The tissue is maintained under normal conditions. It serves as a basis for comparison in evaluating the changes produced by variations in staining the experimental tissue.

Certain metods of chemical analysis require blocking or blanks. Control slides distinguish between true positive reactions and falsely positive non-specific reactions.

Copper containing proteins

Brown black sites point to copper containing proteins. It is electron dense. The treatment of plant tissue growing on copper mine spots consists of potassium ferricyanide, diamino-benzidine and osmium tetroxide in succession.

Corallin

It is rosolic acid. It is an excellent reagent for callose. A saturated solution in 4 per cent aqueous sodium carbonate gives a light red band of callose. Best results are obtained with fresh solution as well as material.

Cotton blue versus carmine

Pollen tubes passing through the style take bright blue staining with lacto-phenol cotton blue. It is superior to carmine. The method minimises the time, dissolves the pigments, clarifies the tissues, removes the fixative and facilitates the penetration of the stain.

Cresyl blue brilliant

The plant chromosomes of the interphase nuclei show red-brown to nearly black stain, the nucleoli and cytoplasm do not stain at all.

Cresyl echt violet

It stains Nissl granules from blue to violet; while basal ground substance takes very pale blue colour.

Cytochemical hybridisation

It is useful to detect and localise specific nucleic acid sequences in microscopic preparations of chromosomes, cells or tissues. In order to make nucleic acid hybrids detectable, they have to be labelled with a light absorbing or fluorescent chromophore, a cytochemically detectable enzyme or an antigen or a hapten. The introduction of the radioactive label must not interfere with the high sensitivity and specificity of the hydridisation reaction. This will form a new chapter or an area in cytochemistry.

Cytochemistry

The chemicals are identified by specific microchemical reactions in a cell. Its quantitative determination will reflect the cell behaviour.

Cytochrome oxidase

The enzyme is very sensitive. It oxidises a mixture of α-naphthol and dimethylparaphenylenediamine to indophenol. The cytochrome C is inevitable for the reaction. It is known as NADI test. The site of enzyme activity is intense blue. Control is run by the complete reaction mixture plus sodium azide. Without cytochrome oxidase C, lipids are positive hence the danger. As for Xanthium, the central zone, pith meristem and leaf primordia were coloured. In Pharbitis the outermost layer of the shoot apex took stain. In the vegetative shoot apex the highest amount of enzyme activity was in the subterminal zone.

Cytofluorometry

It is not the absorption of light which is measured but the amount of fluorescent light emitted when visible or UV light is passed through a stained slide. This technique is highly senstive. Fluorescence is directly proportional to the exciting light intensity which tends to be negligible for measurement.

It is a tool for applying quantitative cytochemistry to cellular composition and perhaps the most important basis at the level of the intact cell and its structure. It is based on the visual observation of qualitative chemical criteria in correlation with the resolved structure of cells and tissues. It combines the optical measuring techniques of anisotropy, diffraction, scattered light absorption photometry, interferometry forming the pivot of flurometry.

Cytophotometer

It records the transmittance of light passing through a histochemical preparation reflected by the mirror on a light dependent resistor connected to a microammeter. Its reading is the direct function of light transmission through the stained slide. It can measure the light absorption of an object of 10 μm in size. It can be assembled from the indigenous materials. The diameter of even the smallest object whose absorption can be precisely measured in visible spectrum by the improved design is 1.5 μm. It obeys the standard laws of light absorption. Its performance is equivalent to market paraphernalia and the calibration checks show that the stability of the assembly is excellent.

DAB for peroxidase

Dark black deposits of the osmicated polymer point towards the sites of peroxidase. The test medium is 3,3'-diamino-benzidine (DAB) with hydrogen peroxide. Paraformaldehyde-gluteraldehyde fixative preserves the ultrastructure also.

DAPI

It is 4',6-diamidino-2-phenylindole. By the DNA-specific fluorochrome DAPI, the nuclei of the second pollen grain mitosis and sperm differentiation are rapid and reliable. Triton X-100 incorporated in the staining solution allows a free access to nuclear DNA. The nuclear status of the pollen grains becomes immediately clear.

Darrow red

Only element in the nucleoplasm stained red is the nucleolus. It is a new basic dye which is is metachromatic. It is excellent counterstain for Luxol fast blue. The stain is stable and specific.

DDD reaction

The interaction of sulphydryls with disulphide groups constitutes the most specific reaction of sulphydryls. DDD reacts with thiols of proteins by exchanging hydrogen protein -SH. The sulphide reagent is split one of which attaches to a protein, the other half is reduced to 2-hydroxy-6-naphthyl sulphide which is subsequently extracted with ether together with any remaining DDD. The colourless naphthol is converted in a second procedural step into a coloured azo dye by a number of diazotized compounds (Barrnett and Seligman, 1952). The site of sulphydryl groups will appear blue with diazo blue B but red with diazo red RC.

Dehydrogenases

It invariably requires the presence of NAD or NADP in order to catalyse a particular reaction. However SDH is an exception to the above. The detection of dehyrogenase activity requires the use of tetrazolium compounds. The sites of enzyme activity appear as blue-black deposits. However its specificity needs a signal of caution.

DGD

Diethylene glycol distearate (DGD) and cellulose caprate resin in 4:1 form a convenient medium for trimming, shaping and to form consistent flat sections. It is a superior embedding medium than paraffin.

Diaphanol

The chloridioxide-acetic acid is an excellent softening as well as bleaching fluid for hard palm and cycad stems, rootstocks of ferns, dark and old barks of neem, and even necrotic woods. It is extremely effective in decolorizing tannins and resin deposits.

Diazo methylene violet

The reaction product with total proteins is black to deep grey violet. Staining is marvellously reproducible for cytophotometry.

Diazotized sulphanilic acid

Phenols colour yellow to light orange. Sulphanilic acid 0.3 g is dissolved in 2mL or 10% aqueous Na_2CO_3 solution.

Dichotomous key for histochemical detection

With a large number of methods and dyes, it has been possible to frame a classical analytic chemical key based upon the main groups of compounds located in a cell. These form groups by themselves. Such five groups are identified: soluble substances, carbohydrates, lipids, proteins and nucleic acids. In any key, each statement of a couplet (DNFB versus fast green for proteins) is termed a lead and each lead is identified by a letter or figure. In the present era, before proceeding to qualify a histochemical detection, such a key will apprehend the procedure to be adopted. Such keys are more prevalent in the powder or in detecting the binomial name of the plants.

Differentiation

It is a process/procedure to overstain at specific loci followed by a careful removal of the excess stain or by progressively gradual increase of the stain intensity. It demands utmost skill. The tissue is deliberately overstained and then placed in water or alcohol or acidified alcohol that slowly kremoves the excess dye. This process of controlled removal of a dye is the differentiation.

Dimedone for glycogen

It is a fast blocking agent for aldehyde groups and prevents PAS reaction. It is useful for the control reaction to demonstrate the hisochemical detection of glycogen. This polysaccharide is fairly soluble in water so it is best preserved by alcoholic faxatives.

Dinitrofluorobenzene (DNFB)

It reacts with α-amino groups and ε-amino groups of lysine, phenolic -OH groups of tyrosine, imidazole groups of histidine and sulphydryl groups. This stain is specific and stoichiometric for total proteins and results in a yellow magenta colour with a peak absorption at 400 nm. However histidine, tyrosine and -SH and NH_2 groups stain deep magenta if K acid is utilised but deep red with H acid. Various reactive groups may be selectively blocked by deamination with sodium nitrite in acetic acid or acetylation with acetic anhydride in pyridine.

Displacement

It is a sequential loss or diffusion of the stain or its replacement by dyes dissolved in ethyl cellosolve. Decreasing the concentration of the dyes like tartrazine, phloxine amido black, fast green, alcian blue simply increases the time necessary for displacement and the resulting stain has the same intensity as before.

Distrene 80

It is a polystyrene of inert molecules with high molecular weight and prevents acidification and consequent fading of the dye. It is thus superior to all synthetic and natural resins as a mounting medium. Distrene with tricresylphosphate and xylene tends to be cheaper with constant results.

Dithizone method for zinc

Zinc yeilds red to purple granular positive precipitates. Tissues depict a diffuse yellow to pink background. In frozen sections fat stains from yellow to green.

DMAB-nitrite for tryptophane

Proteins with bound tryptophane form a strong blue pigment. Tryptamine,

3-indoleacetic acid and serotonin also form the blue pigment. DMAB is *p*-dimethylaminobenzaldehyde in a concentrated HCl for 1 minute.

DNFB for proteins

Proteins with amino groups, tyrosine, histidine and sulphydryl groups form reddish purple complex.

DPX mountant

The resinous medium has the advantage over balsam that it does not turn acid or cause fading of stain. The initials stand for its three components-distrene (a polystyrene), dibutyl phthalate- a plasticizer and xylene.

Ebel test

Sites of polyphosphates stain brown black. The regent used is lead nitrate 10% in 0.05 M acetate buffer at pH 4.5.

Embedding for enzymes

Methyl methacrylate is a low temperature embedding medium for enzymes. The tissue blocks polymerised with benzoyl peroxide and N,N-dimethylaniline blocks cut well and the tissue sections show superior acid phosphatase activity.

Eosin-light green

Eosin stains tryptophane containing proteins blood red but light green is bound to amino groups. The reaction has to follow the steps: treat with Cr_2O_3 for half an hour, keep in phosphomolybdic acid, to stain eosin, to light green and dehydrate through isopropyl alcohol grades.

Esterase

It utilises the substrate 5-bromo-4-chloroindoxyl acetate. The indoxyl is oxidised to an idigoid dye. The method is precise but it locates general activity of the enzyme.

Reaction Mixture	*Active sites of reaction products*
1. Indoxyl acetate	bluish granular
2. Naphthol-AS-acetate 1-naphorthyl acetate with Fast blue salt or Fast garnet GBC or Fast blue RR	purple red to brown

Ester wax (Steedman wax)

It is a ribbon embedding medium. It is more translucent and compression at sectioning is less than with paraffin and celloidin. It is completely soluble in a variety of solvents as alcohols, esters, ketones, ethers, hydrocarbons, chlorinated hydrocarbons and essential oils. Thus the popularity of xylene is endangered.

Ethafoam

It is a fine embedding medium — polyethylene foam - in cryostat operations. It cuts easily with a sharp knife and can be shaped by heating. It is compressible but rigid enough to prepare a tight closure when pressed into an opening. It holds its shape and size for months if not exposed to heat or alcohol.

Ethanol I

Ethanol, usually as a 50% or 70 % aqueous solution, has been used frequently as a fixative since it penetrates plant tissues fairly rapidly, appearing to preserve by coagulating many of the proteins by hardening. It is not a successful cytochemical fixative because (a) coagulation of proteins results in inactivation of a number of enzymes; (b) some lipids are solublized; (c) it effects a splitting of lipids from protein to cause an unmasking of lipids and (d) it does not stablize nucleoprotein, which may be dissolved out.

Ethyl alcohol II

It is seldom used as a fixative. It is an important ingredient in formalin-acetic acid-alcohol and Carnoy's fluids. Tissue proteins and nucleic acids coagulate into insoluble precipitates. No dispersion of starch grains occur while storing for indefinite time. Fats and phospholipids are dissolved. Carnoy's formula is a faster and quick so it is less destructive to cytoplasmic inclusions. However it is rigorous.

Euchrysine GGNX

It is a fluorescent dye to localise chromosomes.

Europium acetate

It is a new fluorochrome-simple, specific and reliable for nucleic acids. It forms, europium salts after treatment with 0.05 M thenonyl trifluoroacetone. Their chelate ions fluoresce bright red orange at a very narrow band. It eliminates also non-specific fluorescence.

Evans blue

It is a vital dye. Aqueous 0.1% solution of Evans blue stain dead cells blue while living cells containing a nucleus and cytoplasm remain without a stain.

Experimental cyanin red

It stains nucleic acids bright red and mucopolysaccharides dull red. Solution (0.5 and 1.0 %) in distilled water is stable in light and at room temperature. It keeps for 4-6 months. Controls are run by enzymatic and acid extractions. It is a astrazone type of cyanine dye.

FASGA

The starch grains fluoresce greenish yellow under fluorescent microscope. But in light microscope the cellulose cell walls stain blue while the lignified and cutinised walls stain red. It is an abbreviation of Spanish names of fuesina, alcian blue, safranina, glicerina and aqua. It forms FASGA.

Fast blue BB

Phenols are located by red to brown colour in roots or stems of *Pinus* sp. *Acacia, Prosopis* etc.

Fast green

At pH 8.00 only histones (high in basic amino acids) still remain below their isoelectric point. By placing the sections at pH 8.00 in acid dye like fast green the only proteins binding with the dye are the total histones. The reaction is quite specific and the colour developed is proportional to the amount of histone and protamines present. The dye-histone complex is green (Fifford and Dengler 966. Amer. J. Bot. 53:1125-1152).

Father Vincent Raspail

He was the architect and founder of microscopic Botanical histochemistry. His research paper met with a tragic fate. It was rejected by a committee of 3 experts. The physiologist was ignorant of chemistry, the chemist of microscopy and the botanist of both. The procedure that requires a synthesis of these 3 fields got now deeply rooted and a fresh momentum.

Female gamete or ovum or oosphere or the egg

In the juvenile egg, the density of DNA in the nucleus becomes highly feeble so that Feulgen staining fails to detect it. The reduction in stainability during maturation of the nucleus is ascribed to the increasing size and consequent dilution, or chemical transformation of chromosomal DNA, stretching of DNA molecule etc. However it reflects the existence of unusual DNA metabolism in the egg cell. At a critical moment, it ceases to divide and loses the ability to replicate DNA. The increasing volume of the nucleus, consequent dilution of DNA and possible lampbrush configuaration of the chromosomes reduce the concentration of DNA.

Ferric chloride test

Phenols yield a lively green colour by 2% ferric chloride in ethanol. It is a specific test under light, SE and TE microscopy. Certain catechol derivatives are characterised by an intense cherry red colour or precipitate.

Ferric-ferri-cyanide test

Proteins containing—SH groups become bright blue coloured. The test solution contains ferric sulphate and potasium ferricyanide. A brief rinsing in alkaline alcohol extremely reduces the blue staining of the background, emphasizing the strongly positive -SH containing elements. Control by saturated phenyl $HgCl_2$ gives no colour.

Ferric sulphate

A blue precipitate indicates the presence of tannins. Ferric choloride in 0.1 N HCl gives intense blue line.

Feulgen reaction after hydrolysis at room temperature

Hydrolysis for 15 minutes in 5 N HCl or 10 min hydrolysis with 5 N HNO_3 at room temperature gave as good staining results as 4-6 min hydrolysis in 1 N HCl at 60°C. DNA delineated magenta colour.

Feulgen technique (test)

It is a staining method with Schiff's reagent. It gives reddish magenta or purple colour where there is deoxyribonucleic acid e.g. chromosomes. The Feulgen test has to be strictly controlled since any free aldehyde in the tissue will react with decolourised para-rosanilin and yields a positive colour. The colour produced is directly proportional to the amount of DNA present and can be measured on the cytophotometer.

Filiform apparatus

These are number of striations converging towards the apex of the synergid cell. It is a highly convoluted extension of the micropylar or upper portion of the synergid wall. It consists of high PAS—positive material. Under UV microscope it is polysaccharidic. The chief component is hemicellulose. Proteins are frequent. In *Aquilegia* it is devoid of cellulose. In cotton it is calcium-rich.

Fixation

It aims at killing cells in such an ardous way that prevents subsequent changes through metabolism or decay. It prevents any disruption or distortion of cell and tissue organisation which ultimately affects the morphology of the cell. Whether the mechanism of the fixative is rational or not, it must give good preparations. Maheshwari's fluid is found superior to Navashin and CRAF fixatives for plant tissues. It preserves chemically or fixes the tissue. Specimen pieces are immersed in at least twenty times their own volumes and shaken vigorously.

Floral apex

The terminal and axillalry buds ultimately develop into flowers. It lays down bulges on either side or the elongating apex. This is coupled with the increase in the ratio of ascorbic acid/histone and DNA/histone. It leads to an increase in total RNA content per cell by pyronin detection as the inhibitory action of histones has ceased to work. It permits a flower to develop.

Flow cytometry or Fluorescence cytophotometry

It measures the flourescence of a large number of single cells in a fluid stream. Feulgen stained nuclei are kept in aqueous suspension and fed through the objective by means of a longitudinal fluid stream carrying the stained nuclei along the optical axis of the microscope at a flow rate of 100 to 1000 cells per second. While passing the nuclei are excited and the total fluorescence emission of each single nucleus is recorded and classified by an impulse height discriminator.

Flurescein diacetate

It has been widely used as a vital dye to determine the viability of pollen grains, cultured plant callus cells and isolated protoplasts. The enzymic cleavage of the acetate accumulates fluoresce in and the emerging bright greenish yellow fluorescence checks the viability of cells. The test is primarily one for the integrity of the plasmalemma of the generative cells lying within the plasmalemma of the vegetative cell. It effectively assesses pollen quality. Living cells exclusively fluoresce while dead cells do not at all.

Fluorescence

Many substances termed fluorechromes absorb UV radiation or white light and emit or throw a light of longer wavelength, e.g. quinine sulphate absorbes UV light and fluorescence blue and thus works as optical brightners. The emission stops immediately the radiation source is cut off. Thus fluorometry has opened up new vistas in quantitative cytochemistry and the understanding of the molecular organisation of the cell. It has been possible to precisely locate the origin, transference and role of sporophytic proteins by it.

Flow Fluorometer—Flow Microspectrofluorometer

It is a device to perform a high speed quantitative karyotyping by flow microfluorometry. The technique involves isolation of metaphase chromosomes from cells, staining with a DNA specific fluorochrom and measuring for stain content at the rate of 10^6/min in a flow microfluorometer. The results tally well with scanning cytophotometry.

Fluorometer (Microspectrofluorometer)

It combines photometric fluorescence microscope with an optical multichannel analyzer. This instrument provides fluorescence emission spectra of biological materials by detecting the entire spectrum simultaneously in real time. These spectra are subsequently recorded and corrected so as to identify the fluorescent reaction products or to test whether fluorescent cytochemical probes bind to the expected substrate within cells.
Ref: Rotz, M-M., Gill, J.E. and Davis, D.T. 1976. A new optical multichannel microspectrofluorometer. J. Histochem. Cytochem. 24: (1) 91-99.

Fluorimeter

It is not prone to distributional errors. The fluorescence emitted is directly proportional to the amount of the dyemetabolite compound present. It has become a well known and a very important tool in the field of quantitative cytochemistry. It yields only relative results: Standard cellular material of known chemical composition is absolutely necessary for calibration.

Fluorochromasia

It is the phenomenon whereby living cells develop an appreciable intracellular fluorescence during incubation in a buffer containing the fluorogenic substrates—like fluorescein buffer containing the fluorogenic substrates—like fluorescein diacetate (FDA). The flourescence is due to florescein (F) which by enzymatic hydrolysis is liberated from the nonfluorescent FDA through cellular estarases. Rhodamine B, and 6G, tetracycline phenosafranin, acridine orange, 3,4 benzpyrene, auramine O form the family of fluorescent vital dyes.

Fluorochromes

When a section is treated with a fluorescent dye and the preparation is illuminated by ultraviolet light, new rays of visible light originate wherever the dye is present. Such chemicals comprise fluorochromes. Different fluorochromes give light of different colours. The object has now become self-luminous. Double strandedness of native DNA is detected by intercalating flourochromes like ethidium bromide or propidium diiodide, its base composition is revealed by Hoechst 33258 or DAPI.

Formal-calcium

It is an excellent fixative for lipids localization. It helps to prevent their movement in aqueous solutions, but does not prevent the lipids from dissolving in organic solvents.

Formaldehyde

It does not coagulate proteins, nucleo-proteins or precipitate DNA from solution. It causes cell shrinkage to an extent of over 60% of the original volume. Lipids are well preserved although some lipid-proteins tend insoluble in lipid solvents. The retention of the activity of some enzymes occurs. Hence it has become a boon in animal and plant cytochemistry, especially in localisation of proteins. Formaline is the most satisfactory fixative for the cytophotometry and UV/microspectrophotometry.

FOTA

It is a feulgen-oxidised tannin azo dye for DNA and proteins in a single section. Feulgen positive material turns magenta bluish red. After immersion in diazotised *o*-dianisidine, tannophilic proteins stain brown to dark salmon red. The nuclear cytoplasmic proteins stain yellowish brown.

Free radical of ascorbic acid

Free radicals are highly reactive as they have one unpaired electron. Ascorbic acid free radical peroxidabse acquires a deep purple colour by an aqueous solution of *o*-dinitrobenzene. A higher AA-FR peroxidase activity is correlated with intense metabolism.

Freeze drying

To overcome the obstacles of chemical procedures an alternate method of rapidly freezing the tissue is freeze drying. The ice crystals are removed by evaporation at low temperature and paraffin infiltrated in vacuum. After freezing in propane or isopentane cooled at -170°C with liquid nitrogen, the tissue can be used for histochemical localisation.

Freeze sectioning

It is a device where fresh living tissues are placed directly in the hole or equilibrated in antifreeze solution before freeze sectioning in a gelatin antifreeze medium. It cuts frozen sections on a microtome. The block holder, tissues and the knife are precooled by compressed CO_2. The sectioning is extremely useful in electron microscope cytochemistry. It permits sectioning of sections which are already cut on a rotary microtome.

Fuchsin acid test

It discriminates between the phenols which are not stained while aromatic amino acids stain well. Proteins become red. It is very soluble inwater but sparingly in alcohol. It stains wood bright magenta. Acid fuchsin has a special affinity for leucoplasts.

Fuchsine reduced

It forms a purple product with aldehydes. It works as a test reagent for the presence of lignin as it has a higher aldehyde components. The reagent is to be carefully decolorised by the addition of strong ammonia drop by drop.

Gahan's fixative

An extremely useful fixative for lipids and especially phospholipids. It consists of calcium chloride 1%; neutral formaldehyde 4%; saturated solution of Reinecke's salt.

Gallocyanin

It is a stain of the oxazine group for total nucleic acids. It is an ideal method where numerous slides need to be stained serially, strictly equivalent and perfectly reproducible results are obtained. Structures containing basophilic compounds take on a bluish black colour. The negtively charged phosphate groups of DNA are stained blue by gallocyanin-chromalum and orange by acridine orange.

Gallocyanin-chrome alum

It is a test for proteins at the sites of histochemical reactions for carboxyls, amino groups tyrosine and histidine. The dye-metal complex is more stable towards light, water and organic solvents than are other cationic dyes.

General esterase

Blue deposits on test sections (embryos of Arachis-ground nut or Hydrilla) indicate the sites of enzyme activity. The test solution contains 5-bromo-4-chloroindoxyl acetate plus potassium ferrocyanide; gum acacia and formaldehyde.

Gersh reaction

Sections are immersed in ice cold cobalt nitrite stain. Quantitative studies by flame photometry show that the stain precipitates potassium in the tissue completely and moreover produces no loss or shift during the staining schedule.

Gibb's reagent

Sites of polyphenol stain blue colour-if forms the ammonium salt of indophenol with 2,6-dichloroquinone-4-chloroimine, borate buffer and 5% ammonium hydroxide.

Giemsa stain

It shows three stages of colouration for suberin in plant cork cells: violet in developing or young phellogen and cork cells; green in still living bark cells; blue in advanced stage where the cells are not living. Giemsa is made up of azureosin-methylene blue.

Glucose-6-phosphatase

Brown black deposits indicate the enzyme activity sites but are entirely absent in control reactions. The test medium consists of potassium glucose-6-phosphate with aqueous lead nitrate in tris bufer.

Glucose-6-phosphate dehydrogenase

There are two tests for the above enzyme.

1. Purple blue formazan (NBT) is crystallised at the centres of enzyme activity. Controls remain absolutely negative.
2. Formazan is purple, controls give negative results.

The incubation medium contains glycylglycine, glucose-6-phosphate, NADP and NBT in varying proportions.

Glutaraldehyde

It is used as a fixative 2.5% solution in phosphate or cacodylate buffer for electron microscopy. It is of immense value in enzyme cytochemistry. Post-fixed with osmium tetroxide gives excellent preservation of organelles and ground substance with membrane contrast. It also imparts positive contrast to membranes. A relatively pure sample with low 235/280 nm ultraviolet absorption ratio imparts enzyme recovery.

Glychrogel (Knox)

It is a mounting medium prepared by mixing distilled water (80 mL), glycerine (20 mL), granulated gelatin (3 g) and chrome alum (0.2 g). It soon dries rapidly around the edges of the cover slip.

Glycol methacrylate

It is a plastic based embedding medium. The addition of polytene glycol (PEG) ehnances ribbon formation. When sectioned with glass knives on AO rotary microtome, ribbons are continuous and do not break.

Gomori's aldehyde-fuchsin

Capsules and membranes of the pathogenic fungi, elastic fibres, mucus and granules of the mast cells stain deep purple.

Growing apex

The vegetative apex has low RNA conent, low ascorbic acid content and low DNA, but a higher conent of histones as detected by histochemical techniques. This indicates that it is passive in view of the metabolic reactions going on in adjacent tissues. With the arrival of the floral stimulus-photophytohormone ascorbic acid, floral induction or reproductive patterns arise on the apex.

Gum adhesive for sections

Its composition is shown below:

Distilled water	98.00 mL
Gum arabic	1.00 g

Potassium dichromate small crystal or menthol

It is an excellent adhesive for sections on the slide. It saves time. Stir well before use. But treatment with acids will wash off the sections.

Hair spray fixative

It is a pleasant surprise to use comercial hair sprays as fixatives for hematological cytochemistry. The reaction product responded to carbohydrates for PAS, lipids by Sudan black, nucleic acids by methyl green pyronin, and enzymes like peroxidase and alkaline phosphatase. The reaction is the replica of histochemical fixatives in use. But the hair spray contents are partly polyvinylpyrollidone and dilute alcohol and some unidentified remains.

HEMA

It is a par excellence embedding medium. It is entirely free of two major drawbacks of polymerization and extreme sensitivity.

Heparin

When heparin is added to Bismark brown, the colour reaction changes from orange to yellow, with Azure A, the change is from blue to red, with cresyl blue from light blue to purple, with Nile blue sulphate from green to reddish blue, with neutral red from red to orange, with pyronin and acriflavin, the solution loses its influence. These colour changes confirm the specificity of the reaction and are termed metachromatic reactions.

Herr's fluid

Ovules/shoot apices, nucule and globule of *Chara* cleared in benzyl benzoate- 4 1/2 fluid can be permanently mounted in diaphane, glycerine jelly, karo syrup, synthetic resins, canada balsam, gum dammar. Herr's fluid is composed of lactic acid. Chloral hydrate, phenol, clove oil and xylene (2:2:2:2:1, by weight). A complete investigation from algae to angiosperms by employing the above clearing technique has been possible. The lone limitation is its temporary nature making oil immersion observations difficult, but photographs and drawings help to keep records. It is an ideal fluid for instant detection.

Heteroauxin

It is an active metabolite which is stained cherry-red by ferric ammonium sulphate in sulphuric acid water solution.

Heterocyst

In the filaments of certain blue green algae, a unique cell interrupts the lineage and is known as heterocyst. Presence of ascorbic acid was detected cytochemically by silver nitrate reagent kept in an amber-coloured bottle (Chayen, 1951). The threads were rinsed with sodium thiosulphate. The cells showed colloidal dark brown deposits of silver.

Hexamine silver reaction

The aldehyde of DNA involving hydrolysis in molar citric acid or N HCl at 60°C followed by hexamine silver reaction and treatment with Schiff's reagent aquire beautiful magenta clour.

Hibicin

The dye gives a bright stable red colour to neutral mucin. It can be used as a specific stain for mucin in paraffin sections. Acid mucin and glycogen do not take this stain. It is a colouring material obtained from the flowers of *Hibiscus sabdariffa.*

Histidine

Sites cotaining histidine are stained in various shades of green and yellow with diazotised sulphanilic acid in conc. HCl.

Histochemistry

It is only a technique for the distribution of specific chemical substances or (sites) residence of chemical activity in cells, tissues, tissue systems and even organs by specific staining methods. It expresses the chemical data in morphological terms either directly as amount per cell or indirectly by correlating chemical change with morphological expressions. These methods are especially useful when heterogeneity of the tissues perchudes or masks the use of conventional biochemical techniques. With appropriate controls, a given chemical substance can be quantified to yield data on its concentration. It translates its activities in relation to form and function.

Histoenzymology

It is the *in situ* localization of enzymes in tissues and cells. It is important to use sectioning or preparation procedures which do not alter the distribution, activity and site of an enzyme. The reaction product should be microcrystalline and amorphous. In addition to the appropriate substrate, a suitable buffer is essential to stablise the pH, an enzyme activator and also a compound to stablise the primary product of reaction.

Histones

They are basic proteins stained by two methods (i) fast green at pH 8.0- (Alfert and Geschwind, 1953) and (ii) ammoniacal silver nitrate (Black and Ansley, 1964). The trends of histone levels in microspores of *limnophyton* are identical in both the above reactions. It directs the same localization by two techniques based on different principles.

Hoepfner-Vorsatz reagent

The action of nitrous acid on a phenol produces a nitroso derivative. It forms a coloured salt on addition of the base like 2 N NaOH.

Hydrolysis

Hydrolysis is defined as reaction of a large molecule with water to form smaller molecules. In Feulgen staining mild acid hydrolysis breaks the purine-sugar glycoside linkage in the DNA and the aldehyde groups are exposed. With gradual increasing time of hydrolysis the ultra-violet absorption of purines and pyrimidines decrease and Feulgen dye DNA complex reaches a peak magenta colour. In contrast with DNA, RNA is Feulgen negative. However, in cytochemistry of DNA, the time of hydrolysis is a critical and deciding step. 1 N HCl at 60°C or 5 N HCl at 20°C serve the purpose well. Murgatroyed (1968) preferred Helley's sublimate formalin fixative for DNA because it gives wide latitude in hydrolysis time for which optimum staining is maintained hence results remain reproducible.

3–Hydroxy-2-naphthaldehyde diazo dye

When the highly coloured diazo blue B is used, the end colour for total proteins ranges from intense red to blue. When the diazo red KC is used only a brilliant red product is obtained. The terminal amino groups of proteins combine with the aldehyde containing compound-3 hydroxy-2-napthaldehyde to form a colourless end product which is made coloured by the azo dye.

Hyrax

It is a mounting medium meant exclusively for diatomaceous earth. It is a resin obtained from napthalene. Its refractive idex is 1.8. When a little hot, it displaces air from the objects. It is neither an acid nor a base, hence slides in a cabinet remain for 5-6 years unaltered.

IKI-H_2SO_4 method for plant cell wall

Cell walls containing cellulose stain dark blue. Even in orange to yellow lignified walls, the dark blue cellulose linings are detected. The middle lamella will develop green. Highly lignified elements and wood stain well.

Immuno-histochemistry of proteins

The FTIC (fluoresce in isothiocyanate) acts an antigen and protein molecules activity reflect as antibody. Both will combine in a cell by a covalent bond and form a product-conjugate. When blue exciting light is used, a yellow or orange barrier filter is necessary and the autofluorescence appears green.

Indoxyl esterase

The enzyme imparts light blue deposits by the substrate 4-chloro-5-bromoindoxyl acetate. The following criteria prove its specificity.

1. zero order breakdown of the substrate
2. first order capture reaction has a high velocity constant
3. non-diffusibility of the deposits.

Insoluble polysaccharides

An intense red purple product is formed by using 3-methyl 2-benzothiazolinone hydrazone for total polysaccharides.

Integrating microdensitometer

This instrument scans selected areas of a specimen, transmission is measured by a photomultiplier and associated integrating circuits produce a voltage inversely dependent on the total light energy passing through the entire scanning area. A measure of absorption is obtained by comparing readings with the specimen in place with those from a neighbouring blank field. Selection of scanning wavelength is by a gaded filter giving a band width of 14 mμ to 50 per cent peak tranmission through the range of 400 to 700 mμ. The results are superbly reproducible.

Insulin

It occurs in the storage organs of many monocotyledonous plants, dahlia tubers and members of Compositae and Campanulaceae. It is readily precipitated in the form of sphaeracrystals if the tissues are fixed in ethanol. If the precipitated inulin is immersed for 2-4 minutes in alcoholic phloroglucin, a red brown colloidal form is obtained.

Iron

The metal is present in the ferric form. In the presence of ferrocyanide ions, the stain Prussian blue is immediately precipitated in blue iron-product after several hours. Equal parts of aqueous 2% sodium ferrocyanide and 2% HCl are mixed immediately and stained with Prussian blue. It is known as Pearl's reaction.

Iron-pyrogallic acid

It stains the filamentous structures-nucleolonema as fine black particles or threads inside the nucleus. The material is to be fixed in FAA. Silver impregnation technique gives bright results.

Jenus green B

It is a basic dye. Mitochondria stain blue in living cells only, but lose their stain in anaerobic conditions. Dead cells are uniformly stained blue green throughout. Thus the particle identity is inherent in the procedure.

Journals on histochemistry

Acta Histochemica; Analytical Biochemistry; Analytical Chemistry; Annales d'histohimie; Archives of Biochemistry and Biophysics; Biochimica et Biophysica Acta; Comptus rendus, serie chimique et series physiologique; Experimental cell research; Journal of histochemistry and cytochemistry; Journal of cell biology; Stain Technology; Quarterly Journal of Microscopical Science; Histochemistry (formerly Histochemie); Caryologia; Cytologia; Protoplasma; Chromosoma; The Nucleus; Histochemical Journal: Journal of Microscopy (formerly Journal of the Royal Microscopical Society).

Kallichrom

It is a combination of cresyl violet and auromin. The saturated solution takes a cherry red colour. It must not have a yellowish or purplish tinge. In plant tissues, nuclei become dark blue, xylem and fibres green and cellulose red to pink.

Karo

It is a corn syrup containing maltose, dextrose and dextrin. It is extremely inert. It can be used to advantage in mounting algae, pollen grains and entire insects. None of the sugars crystallise hence karo is superior to glycerine jelly.

Kino gum by triple stain

The blood red or ruby-coloured gum exudes from the bark of *Pterocarpus marsupium* Roxb. The kino gum consists of tannin or tannic acid (25-80%), pectin and carbohydrates. Ferric ammonium sulphate, ruthenium red and PAS reactions respectively are used to test these constituents.

Latex

It is a source of natural rubber, medicinally important alkaloids and proteolytic enzymes. Laticifers are internal secretory systems which contain latex. The tissue is stained with oil red-o imparting reddish orange colour while neutral red imparts dark red colour under bright field microscopy.

Lead acetate reaction

Plant tissue pieces are fixed in neutral lead acetate solution. Sections are stained with a freshly prepared solution of H_2S in distilled water or with ammonium sulphide, alcohol. The marked result is a black precipitate in the nucleoli but none in the chromosomes or chromatin of the interphase nuclei.

Lead tetracetate

It promptly reacts in glacial acetic acid to attack α-hydroxycarboxylic linkage in plant tissues.

Leucine aminopeptidase

The ensyme activity is measured by using diazo blue B and the substrate is L-leucyl β-napthylamide yielding a red azo dye. By cupric sulphate there is a change of colour from red to blue.

Leucobasic fuchsin

Schiff (1866) introduced a reagent employing rosaaniline SO_2 for *in vitro* detection of aldehyde. Feulgen and Rossenbeck (1924) applied this regent for *in vitro* localization of DNA following staining of acid hydrolysed tissues. In the aldehyde reaction a peagent is prepared in a leucostate with basic fuchsin, containing SO_2. It decolourises the dye in a leucoform by potassium metabisulphite, thionyl chloride, sodium sulphite or directly bubbling SO_2, folowed by shaking with activated charcoal reduces the dye to a water like solution. DNA is coloured reddish purple or magenta colour.

Leucyl naphthylamidase

Sites of the enzyme activity become bright red or purplish blue. The incubation medium is either L-leucyl-β naphthylamide or L-leucyl-4 methoxy-2-naphthylamide. Control is run by the ommission of the substrate from the test solution.

Lewitsky's fixative

Lipids are made insoluble in the above fixative composed of chromic acid and formalin. The chromate ions make them partially insoluble. Formal-calcium which fixes the lipids, is infiltrated with carbowax or polythylene glycols. Lewitsky's fluid is good for phospholipids since it permits paraffin infiltration.

Lignin

It is a macromolecule deposited along with the cellulose in the cell walls of woody plants. It gives added strength. It is stained bright yellow by aniline sulphate, red by phloroglucin, bright red by chlorine sulphite and purplish red by Schiff's reagent.

Lipids

They are extremely difficult substances to work with histochemically as organic solvents remove lipids as well as non-lipid substances. The physical methods employ Sudan stains, Nile blue and the vital dyes. On the other hand the chemical reactions employ haematin, orange G and per-acetic acid-Schiff for lipid detection.

Losses in DNA

Unifixed tissues proved a huge loss in DNA using tissue homogenates or stained preparations by Feulgen reaction. Fixation and acid hydrolysis definitely reduced the amount of DNA. Fomalin and alcohol-formalin gave the smallest or negligible losses of DNA.

Luxol fast blue MBS

It is insoluble in water but soluble in alcohol and plar organic liquids. Mitochondria are stained green, selectively and clearly. If counterstained with phloxine, it gives a better colour contrast. The stain is durable, highly specific, simple and gives a better colour contrast. Luxol fast blue MBS (du Pont) stains phospholipds dark blue to purplish.

Luxol fast yellow TN

It stains cytoplasm brilliant yellow. Thus it works as an effective counter-stain for nuclear stains like celestin blue, pontacyl blue black SX and P violet 6R. Developing fibres take a decisive stain.

Magnesium and calcium phytates

They are organic phosphates. Tissue sections are treated with 20-30% $FeCl_3$ or basic dyes like pyronin B, safranin, Bismarck brown or basic fuchsin. These phytates of calcium and magnesium are strongly stained with respective dyes. These form tubiform or fingerlike outgrowth. It is a specific detection procedure for calcium and magnesium.

Maheshwari's fluid

It contains

Ethyl alcohol 50%	90 mL
Glacial acetic acid	5 mL
Commercial formalin	5 mL

(40% formaldehyde) —It is as a universal and ideal fixative and excellently balances the opposing tendencies of the three regents to cause swelling and shrinkage during fixation. It gives the best results with parafin embedding. Yakovlev, M.S. and Yoffe, M.D. (1957) working on some peculiar features in the embryogenesis of *Paeonia* named the fluid first after Professor P. Maheshwari.

Malate dehydrogenase

It utilizes sodium malate in the incubation medium to indicate the activity by the formation of fine formazan.

Manganese

The periodate—tetrabase test in forzen woody stem sections of apple bark produces a blue colour where Mn is present. Colour intensity is directly proportional to the amount of Mn present. N,N-dimethylamiline is the tetrabase while precipitates are of its oxidation products.

Maule's test

Lignified vascular tissues become dark brown due to MnO_2 deposition by treating the sections with 1% w/v $KMnO_4$ for 1 minute only. Weak or developing lignified tissues do not contain syringyl units, hence the colour is faint. Thus syringyl units are present in the early stages of development but not in detectable amount.

Maxilon blue RL

It is a metachromatic, monoazo, cationic dye with different alipathic or aromatic raicals when applied to tissue sections, it stains nuclei blue. Acid mucopolysaccharide elements stain red to violet.

Megasporocytes

Meiosis in potato is studied by ovule squash technique. The enzymatic maceration followed by squashing in acetocarmine or aceto orcein, megaspore mother cell (female reproductive cells in the ovule) contained chromosomes nicely spread and stained bright red.

Mercaptide formation

The sulphydryl proteins take red to pink hue. The reaction is extremely specific. The reagent is 1-(4-chloromercuriphenylazo)- naphthol-2 (mercury orange).

Mercuric bromo-phenol blue

The proteins are stained intensely blue. Staining is stoichiometric and folows Lambert-Bear law over wide ranges of concentration. To void loss of the bound dye ethanol serves better than tertiary butyl alcohol. Proteinaceous material similar to p-protein is reported in the phloem of the rachis of the fern *Pteris longifolia*. The solution consists of 1% $HgCl_2$ and 0.05% bromophenol blue in 2% aqueous acetic acid.

Mercuric chloride

It is a poor but fairly rapid fixative. Nucleic acids and proteins coagulate, lipids undergo plasmal reaction, and enzyme activities are inhibited, but organelles of a cell remain well preserved. It is a powerful coagulant of nucleic acids and proteins.

Mercurochrome

It can be used for light (reaction time 48 hours) or (reaction time 1 hour) fluorescence microscopy. If excited by blue or ultraviolet light, bright green fluorescence at sites of sulphydryl proteins but appear red to pink in light microscopy.

Mercury orange

It is highly critically tested means to visualise sulphhydryl groups. Benett and Watts (1958) emplyed it for quantitative determination. - SH groups react with mercuric compounds to form coloured mercaptides. The two forms of sulphur are identified by it. The method was first designed by Hellermann and stained by Barka and Anderson (1963). The mounting medium should match with the refractive idex of the specimen. Care should be taken not to dilute the mounting medium with an excess of xylene on the slide. Sulphydryl group sites stain blue to pink to red.

Metachromatic

It is the colouring of different tissue constitutents in different colours by a single dye. A substrate which causes a metachromatic change is called a chromotrope and it carries a change opposite to that of the dye colour. A well known metachromatic dye is Toluidine Blue 0- a thiazine, or aniline blue. They impart colour other than the fundamental colour constituent.

Methylal

It provides a method for embedding plant tissues without dehydration. It is a quick saving time device. The finer details of cell structure are well preserved though an important step of dehydration is skillfully omitted.

Methyl green alone

Sites only of the DNA stain green. It is essential to immerse the sections in 0.1 N HCl for a 15 min at room temperature to release or free the DNA from its binding sites with histones or basic proteins.

Methyl green-methyl-violet union

MG-MV stains cytoplasm light green, nuclei and chromatin dark green while collagenous as well as elastic connective tissue blue violet. In plant cells cortical tissues showed colour contrast.

Methyl green pyronin Y stain

It stains chromatin clear green, nucleoli bright red and cytoplasmic RNA brilliant red-pink. The stain is prepared by dissolving 0.15 g of methyl green in 100 mL of 0.1 M acetate buffer (pH 4.7) or HCl-sodium acetate buffer extracting it with chloroform until the latter ceased giving a violet colour and adding 0.25 g of pyronin to it. The specificity is ascertained by treating the sections prior to staining with RNase and DNase or extraction with perchloric acid.

Methyl methacrylate

It is an artificial plastic and used as a mounting medium. It dries more quickly than the common resins and does not at all show discolaration on aging. No harmful action on strains has been observed. It is insoluble in ethyl alcohol, amyl acetate, xylene but readily dissolves in chloroform and dioxan.

Methylene blue

It stains the mitochondria, cytoplasmic vacuoles and other granules in living cells blue. The dye readily penetrates the nucleus and stains deep blue the nucleolus folowed by coagulation. Dead cells stain diffusely.

Methylene violet

It remarkably brings out red azurophil granules of mono and lymphocytes when used with eosinated thiazins by zinc-alkali chlorate hydrolysis. The pure dye is insoluble in water but dissolves in the presence of other thiazine dyes. Such haematological stains have been developed empirically.

Microincineration

The section mounted on a slide is inserted inside a muffle furnace at 650°C. It effectively destroyes the cytoplasm and nuclei leaving only an ash of non-volatile minerals. With the microscope with dark-field illumination oly the mineral deposits shine. Iron appears yellow to deep red, phosphorus, magnesium, calcium, potassium and sodium appear white and silicon is crystalline. If the carbon or ash is brownish or black, the slide is faulty.

Microphotometry

It utilises reflected light for quantitative measurements in chemical identification of (coloured) stained compounds. It provides absorption curves of intact cells, the amount of pigment present, the intensity of a particular reaction, the rate of colour production and the reactions involving one or several coloured products etc.

Microscopic histochemistry

Its architect was a French botanist-Raspail, F.V. in 1930; his first research paper was found extremely poor. The reason for rejection was that the physiologist was ignorant of chemistry, the chemist of microscopy and the botanist of both (Baker, 1943). The bugle of the trio opened up a new frontier.

Millon's reaction

In this reaction, nitrous acid and mercuric ions attack tyrosine and tryptophan residues to form nitrosomercurial derivatives of these amino acids. The correlation between the amount of Millon chromophore and model proteins at 280 nm in various cell types is linear. The specificity of staining is tested without nitrite or by previous blocking the phenolic hydroxyl groups with DNFB. Proteins containing tyrosine appear red to pink and orange.

Microtomy

The standard procedure involves the following sequential steps:

1. Fixation in a chemical fluid that can preserve cell structure.
2. Tissue dehydration to remove water from the cell by alcohol.
3. In filtration by careful replacement of xylol by paraffin.
4. Sectioning at uniform thickness to permit comparison.
5. Staining by specific and stoichometric testing and ultimate differentiation at peak level.
6 Covering by a coverslip of uniform thickness on permanent mounting medium. It should not diffuse stain.

Each step needs meticulous care if a satisfactory histochemical preparation is to be obtained.

Microspectrophotometer

An extremely small amount of the dye-metabolite complex used in histochemical methods presents specific problems when working with colour reactions. The law of Lambert and Beer about the light absorption provides the basis for all photometric observations on solutions as well as stained tissue components. Such measurements are comparative and not absolute. Absorbance is linearly related to concentration. First by means of an amplification control, the indicator is set on 100% deflection. Now the light is converted into a photo-current and can be measured after amplification on a microammeter. Now the control slide is replaced by the stained one and it is noted how much deflection is reduced. The readings are affected by light scattering, stray beams, too large a bandwidth, malalignment, shape and thickness of the sections and irregular distribution of the complex.

Mithramycin

It binds relatively specifically to double-stranded DNA and exhibits highly enhanced fluorescence. It does not show any fluorescence with either RNA or protein. It fluoresces yellow while DAPI (4:6-diamidino-2-phenylindole) blue. The fluorescence is proportional to the amount of DNA present.

Mitochondria in yeast

Globose as well as thread like mitochondria take lovely green stain by Janus green B at pH 5.6 and 0.01 solution in distilled water. The stain is most specific because its reaction depends upon the release of the enzyme cytochrome C oxidase. Yet lower concentration and short incubation prove superior. Amido black 10B and toluidine blue O are inferior for mitochondrial staining in *Saccharomyces cerevisiae.*

Modified Feulgen reaction

DNA will stain red or reddish purple or magenta. The sections need hydrolysis with 5 N HCl at 20° C for 40 minutes, direct immersion in Schiff's regent for 30-90 minutes followed by three quick changes of 0.5% sodium metabisulphite. This reaction inspite of heavy upheavals has withstood the test of time.

Mucin

It is a secretion product of certain fungi and some flowering plants, Mucicarmine is specific and highly selective for fungal gel which appears uniform light red. With a fast green counterstain it is especially effective for gelationous tissue. It is highly selective for the gel leaving the hyphal walls and cytoplasm unstained.

Murexide

Calcium deposits appear as orange-red spots or areas against a colourless or very faintly pink background. A saturated solution of 1 N NaOH, 6.5 % KCN in distilled water even does not remove the metal from its original site.

Muller's method

Treating the sections by the colloidal iron, followed by staining with the

Feulgen reaction acid carbohydrates are blue and nuclei are magenta. It improves the selectivity of staining and yields specificity.

Myrosinase

Sites of the enzyme activity reveal brownish black hue of lead sulphide (lead sulphate) medium. The test solution is composed of 0.1 M cacodylate buffer with 0.1 M potassium myronate (sinigrin) and lead nitrate. Its control reaction lies in heat-inactivated material. Roots give good results.

Naphrax

It is a synthetic mounting medium which prevents diffusion of the stain. It is a stable resin with a high refractive index of 1.7. Addition of 1% of a plasticizer gives good results.

Naphthol blue black

Amidoshwarz stains proteins blue black.

Naphthol yellow S

Arginine rich basic proteins or histones sites stain yellow. Its 0.1% solution in 1% aqueous acetic acid at pH 2.8 is quite specific for anthers and root-shoot apices. It combines quantitatively with the available dibasic amino acid residues of the tissue stain. The two absorption maxima one for DNA and the other of histones are entirely different.

Negative staining

The background takes intense stain but not the section. It is achieved by floating the mounted specimens for 30 S on 0.2–1% silicotungstate or phosphotungstate at pH 7.00. It is more reliable and worthy of results. Heavy metal salts upon drying form a glassy cast of the embedded objects.

Neo-terrazolium test

The sulphydryl proteins are evident by blue hue. The test solution is prepared by dissolving 1 g of sodium cyanide in deionised water with neutral pH. In plant tissues triphenyl tetrazolium at pH 7.2 gives positive colour product.

Neutral buffered formalin

It is a superior fixative for basic nucleoproteins. Tissues remain for several months without any marked change. For enzymes, only 12-24 h at 4°C is the permissible time limit. The reaction does not work if any other fixative is used.

Neutral red

It is a vital dye. It needs no fixation. Neutral red 0.1% in potassium nitrate accumulates only in vacuoles of living cells. It is a general tissue stain for fixed embryological material.

Nevillite V

It is a reliable mounting medium. It is inert, homogeneous, resistant to light, dilute acids and alkalies, like toluene soluble in hydrocarbons and insoluble in water and alcohol. There is extremely less chance of trapping air bubbles in the mount.

Niagara blue 4B

It is an acid azo dye. It has been intensively employed to differentiate vascular tissues of plants. It appears dark to shallow blue.

Nigrosin

It is a strong decolourizing agent and is extremely useful in the staining of sporopollenin fungal spore and algae. It stains the background yellow leaving the actual specimen unstained. Thus it is a negative stain. The ciliary bands, basal bodies and cytoplasmic inclusions become predominant.

Nile blue sulphate

It is a dye of oxazine series. It has 3 components with identical absorption spectra in the visible range representing isomeric forms. Phospholipids stain intense blue by the oxazine base of the dye. Neutral lipids stain red by the oxazone derivative. Free fatty acids colour magenta-red due to a blend of red and blue staining. The histochemical procedure detects phospholipids from the other main lipid classes in tissue sections. Neutral lipids like fats, oils and waxes stain red while phospholipids and free fatty acids stain brilliant blue. It is a water soluble dye. The dye consists of blue oxazin and red oxazone.

Ninhydrin reaction

It is triketohydrindenehydrate. Total proteins develop a reddish purple colour. Deamination and acetylation serve as controls. The reaction involves the production of an aldehyde which is stable and non-diffusible with Schiff's reagent, it yields a precise method for α-amino acids.

Nitrocellulose

Low viscocity nitrocellulose is a good embedding medium. The precaution is that nitrocelluloses are dangerously inflammable. It is dissolved in a mixture of diethyl ether and ethanol. It is hardened by chloroform vapour and the tissue material is oriented.

Nitro-brucine

It is a specific regent for detection of ascorbic acid. Sites of the vitamin C become light bluish black.

Nitron

It forms insoluble salts with nitrates and these can be demonstrated in a tissue with polarised light. The nitrates show doubly refractive zones.

Nitroprusside reaction

A pink colour develops for proteins containing-SH groups. But the stain is likely to fade after 8-10 hours. But it distinguishes islets of cells containing -SH groups very promptly.

Nitroso reaction

Tannins especially catechol acquire acherry-red colour in 10% sodium nitrate, acetic acid and 20% urea.

Nitrous acid test

Tissues containing phenols are indicated by a bright yellow to red colour. Sections are bathed in 10% aqueous sodium nitrate, acetic acid and urea. Addition of NaOH (2 N) hastens the colouration.

NQS test for arginine

Sites of high concentration of arginine rich basic proteins apper red to brown. Aqueous sodium hydroxide dissolves 1,2-napthaquinone-4-sulphonic acid sodium salt. The solution is unstable and must be used quickly.

Non-specific esterases

The enzyme activity is manifested by an intense red product by using naphthol AS-D acetate. But the coupling of fast red violet LB salt with free naphthol ASD is remarkably sensitive to the presence of small amounts of L-ascorbate. Thus there is an interference of L-ascorbic acid with standard histochemical azo dye coupling procedure.

Nuclear fast red

It pinpoints calcium deposits by a red dye lake as in calcium oxalate crystals, calcium carbonte etc. in *Ficus* sp.

Nucleases

Sites of deoxyribonuclease activity are brown black. The incubation medium consists of acetate buffer pH 5.0 with 10 mg DNA, 5 mg acid phosphatase and 0.25 mL of 0.4 M aqueous lead nitrate.

Nucleic acids

The double-strandedness of native DNA is detected by intercalating flourochromes like ethidium bromide or propidium diodide. The negatively charged phosphate groups are stained by gallocyanin-chromalum or acridine orange.

It refuses to reveal non-nuclear DNA. Frozen sections and incorporation of (3H) thymidine, its resistance for extraction with weak acid or digestion by ribonuclease and its sensitivity to deoxyribonuclease helped to demonstrate cytoplasmic DNA. In plants and animals it occurs in both chromosomes where the sugar is deoxyribose and in cytoplasm and nuclei where the sugar is ribose.

Oil blue NA

It is an oil soluble dye. It specifically stains rubber deposits in guayule plant bright blue. It is more brilliant than that achieved with either Sudan III or IV. Cells or ducts inclusion taking the blue stain is clear rubber. But non-rubber substances are in a large measure removed by the javelle water or NaOCl. The plants that have given excellent results are rabit brush, Russian dandelion, guayule and *Cryptostegia*.

Oil red O

Fats are excellently coloured red or yellow-red depending upon their concentration. Isopropanol is the solvent for oil red O. It is a common stain for hydrophobic compounds such as unsaturated triglycerides, sterol esters and fatty acids. Control is the extraction by pyridine or ether ethanol prior to staining.

Orange-G aniline blue

The method for lipids is quite promising. The tissue needs chromic acid fixative. The orange colour renders phospholipids specific and aniline blue works as a counter stain for nucleo-proteins and proteins. The absorption values do not go in a linear manner.

Orchidaceae

It is a family of monocotyledonous flowering plants. It provides an ideal living material to the study of embryonic processes by histochemical methods. The substances determined are oxidising enzymes like peroxidase, cytochrome oxidase, polyphenol oxidase and dehydrogenases; physiologically active substances like ascorbic acid, sulphydrils and heteroauxin, food substances like reducing sugars, starch, oils, amino acids and proteins and finally cell wall components like pectic substances, cellulose lignin and tannins.

Orthochromatic

A coloured product bears the same colour as that of the stain is orthochromatic.

Osmium IV-ammine (OA) (System I)

Sections showed DNA of magenta colour as in Feulgen-type reaction. The staining showed specificity, integrity and fineness. This newly synthesized inorganic compounds, also stained extracellular polysaccharide moeities with identical light and electron microscopic appearances.

Osmium tetroxide Osmic acid-OsO_4.(System II)

It is a dangerous but best fixative for cell structure. Its vapours cause blindness. It oxidises the double bonds in unsaturated fatty acids of a lipid molecule. Since it dissolves very slowly, the fixative is prepared a day before use. Unsaturated lipids or fats appear black. Addition of 1.5% sucrose to OT acts as an osmo-protectant.

Otan reaction

Cholesterol esters and triglycerate esters stain dark black while unsaturated phospholipids stain orange red by osmium tetroxide-α-naphthylamine treatment. It depicts chromosomal lipids. It is a general purpose reaction used successfully for plant tissues.

Palm pollen

A colchicine-lactose solution rears and stains pollen from buds at anthesis. It stains and preserves bright pink red the chromosomes.

Papain

It ranks first as a clearing agent for whole mounts as cell contents are partly digested by papain. Permanent slides of plant tissues can be prepared after staining with 1% saffranin O. Preserved materials are preferred for digestion.

Pars reaction

Aldehyde groups formed by the splitting of the 1,2-diglycol linkages take intense pink stain in any plant tissue. This polysaccharide reaction devised by Hotchkiss is specific and well known as PAS reaction as it utilises periodic acid Schiff's reagent. In control it needs the omission of oxidation step. Selective extraction would facilitate the recognition of individual components.

PAS reaction

The reaction is positive with hexose sugars containing glucose, galactose, mannose and fructose. Sections are first treated with periodic acid which oxidises hydroxyl groups to aldehydes which are rendered visible with Schiff's reagent. The cytoplasm remains colourless, the nuclei faint light but the polysaccharides give an intense purplish red. Thus the structure of the plant cell wall shows remarkable clarity.

Pectins

Pectins after easterification are identified by intense red colour. The complex is produced by the reaction of methyl esters of pectin with alkaline hydroxylamine. The hydroxylamic acid yields a red product with ferric chloride. The colour is intensified by treating the plant tissues in a warm solution of methyl alcohol containing 0.5 N HCl. The test is specific and preferable to the use of ruthenium red.

Peracetic acid-Schiff's reaction

Unsaturated lipids stain red. A negative reaction is obtained by treating the sections with 1 mL of bromine in 40 mL of carbon tetrachloride.

Pectic substances

They will appear pink to red when stained with aqueous ruthenium red until the cell walls become bright red.

Perchloric acid extraction

A differential extraction by 1 N perchloric acid step removes the RNA and then in 0.5 N perchloric acid at 70°C removes the DNA. The above procedure is an excellent control for confirming nucleic acid staining by Feulgen method, methol green pyronin or azure B.

Performic acid–Schiff reaction

It delineates unsaturated lipids bright red, but inside, the fat globules remain entirely unstained and clear. Both the reagents are not fat soluble.

Peroxidase

It has offered genuine difficulty of diffusion. However it is localised by 2 methods. Active sites of reaction product precipitates.

1. *p*-Phenylenediamine—Red or brownish red.
2. Benzidine— Blue turns brown

The above reactions are based on the oxidation of benzidine or guaiacol in the presence of peroxidase and hydrogen peroxide.

The problem of diffusion is very acute when intra-cellular localization is desired.

Peroxidase (DAB method)

Fine black deposits confirm the places of peroxidase activity in plant roots. It is entirely absent in controls. The test medium is diaminobenzidine (DAB) with hydrogen peroxide.

Peroxidase by HVA

Brown black precipitates are the sites of peroxidase activity.

Control will not at all be affected. The incubation medium consists of homovanillic acid (HVA) and lead nitrate.

Phenazine methosulphate (PMS)

It facilitates the demonstration of dehydrogenase activity in the absence of endogenous diaphorase.

Phenols

Osmium-potassium iodide (Os-KI) stains dark gray to black phenolic compounds situated in vacuoles of spruce, tobacco, oak, coffee etc. The mixture reacts rapidly with several naturally occurring plant phenols developing solid black precipitates. Phenols containing *o*-dihydroxy units in an aromatic ring function as primary sites of reactivity with the osmium iodide complexes. Os-KI reaction is specific for phenols and omits ascorbic acid reaction by silver nitrate.

Phenosafranin

It is a vital dye. While viewing unfixed or entire preparations under a fluorescent microscope using blue light excitation dead and degenerating cells fluoresce while live cells refuse the entry of the dye.

p-Phenylene diamine

Aldehyde radicals in a variety of compounds are detected by treating tissue sections in aqueous mixture with 1.5% H_2O_2. Structures containing aldehydes are stained bluish violet. The method is extremely sensitive and selective than the leucofuchsin technique presently employed.

Phloroglucinol-HCl

It is a specific test for lignin. Both the secondary and metaxylem are stained deep red-violet while near the cambium staining in new elements is less and on the protoxylem is faint. This test rests on the presence of cinnamaldehyde type side chains in lignin.

Phosphatase and -SH groups by autoradiography

Sections are incubated with 32 P labelled glycerophosphate and calcium nitrate. The enzyme activity is measured as light black precipitates. The method is very simple to quantify with a Geiger Muller counter. Similarly sulphdryl groups can be grain counted with tritiated n-phenylmaleiamide.

Phosphomonoesterase II—Acid phosphatase

The enzyme catalyzes the hydrolysis of *p*-nitrophenyl phosphate at pH 4.8 and liberates inorganic phosphates and *p*-nitrophenol which combine with NaOH to form a yellow coloured complex. Its optical density is directly proportional to the acid phosphatase activity.

Phosphine 3R

It is a fluroscent dye. Under ultraviolet fluorescent microscope lipids apear as a clear milky-silvery white fluorescence. The soaps, fatty acids and sterols will not emit any light.

Photine HV

It is an optical brightener which specifically binds to B-linked carbohydrate polymers. The rate of synthesis of cellulose varies with the intensity of fluorescence under UV microscope. The root hair of *Peperomia* is an excellent object for photine study.

Picric acid

It is a poor histochemical fixative. It inhibits enzyme activities and cell organelles remain highly distorted. Proteins coagulate, nucleic acids partially hydrolysed, and carbohydrats and lipids are unaffected.

Pianese III b

Cotton fibres take green stain and the interspersed fungus mycelia or threads become deep pink. It delimits the extent of infection by the fungus in the fabrics. The pianese stain is martius yellow 0.01 g; malachito green 0.50 g; acid fuchsin 0.10 g, distilled water 50 mL, ethyl alcohol 50 mL.

Pinacyanol

The dye pinaccyanol chloride stains chromosomes and sex chromatin preparations bright blue. The stain is uniform, fade resistant and gives superb details of mitotic and meiotic chromosomes.

Pizzalato method

It is a specific method for developmental studies of oxalate crystals in plants. Stain contains a mixture of 30% H_2O_2 and 5% $AgNO_3$. Crystals show a bright shining silver coating on the surface. Even minute crystals are easily located. Silver reaction product forms a thin, discrete layer on the surface on each crystal mass.

Plasmal reaction

Sites of plasmalogen-(a group of aberrant lipids) give magenta colour product. Fresh or briefly fixed tissues are immersed in mercury chloride and stained with Schiff's reagent. The reaction demonstrates only the effective sites of plasmalogens.

Plasticine

It is a fine embedding medium for making free hand sections to determine the first hand information for the cell contents before microtomy. It is also possible to obtain very thin sections. The shavings of plasticine could then be collected and used again ad infinitum.

Polythylene glycol

It serves as full proof embedding material for semihard seeds difficult for microtomy. Sections temporarily stained by 0.025 to 0.035 thionin offered sharp contrast for photomicrographs.

Polyphenol oxidase

The browning reaction in apples signifies this enzyme activity. They are also known as tyrosinase, catechol ocidase and phenolase. The test solution contains DOPA-D-1,3,4-dihydroxydiphenyl-alanine with sodium diethyldithiocarbomate. Pigmented product is visible at sites of polyphenol oxidase activity but markedly lacking in controls.

Polyphosphate

Sites of polyphosphates stain brown black with lead nitrate (10%) in 0.05 M acetate buffer at 4.5 pH.

Ponceau S

It stains strands in sieve-tube-phloem-quite red or reddish white, and possibly the plasmalemma. Proteins stain bright red. Cell walls do not stain. The best view of main strands and emerging delicate fibres in phloem appears pleasing red in summer.

Pontacyl

It is a textile dye pontamine fast scarlet 4BA and dark green B both stain nucleus green, nucleolus dark-green, connective tissue red and keratin dark green. It acts as a rapid one solution differential stain. Being commercially exploited, it needs great care for result's reproduction.

Mitotic figures are unusually conspicuous as a result of purple colouration by the dye variant pontacyl violet 6R.

Positive staining

When definite components of tissue micro-and macromolecules or elements are pretreated with uranyl acetate or lead citrate, they combine with heavy atoms. It is a positive staining. It stains the sectioned object and not the background.

Postosmicated Hermann's fluid

Mitochondria are shown black against a colourless transparent cytoplasm. They retain their original form. The staining reaction is in Hermann's fluid postosmicated at 34°C and embedded in n-butyl methacrylate.

Potassium dichromate

It is universal fixative. It causes ergastoplasm to coagulate and mitochonria seem to vanish. Dictyosomes are destroyed. Nucleus forms a gigantic mass or an amalgam of coagulated histones. But it fixes lipids very well. Since dichromate fixatives alter the state of tissue metabolites profoundly, they are not used for histochemical investigations. Like chromic acid by dichromate tissue becomes dark. It simulates a tanning process. Like picric acid, their use has nearly perished.

In aqueous solution, it is a strong oxidising agent and softens tissues. It causes shrinkage of the cytoplasm, swelling of mitochondria, renders fat insoluble, does dissolve out DNA and does not coagulate proteins and nucleoproteins.

Potassium permanganate

It is an fixative for obtaining membrane contrast and for electron microscopy. Only DNA is retained but RNA and histones are removed, monosaccharides and lipids are lost and phospholipids are unmasked.

Proteases

Its activity is determined by a novel method by employing the use of photographic colour film as a substrate. The three pigment layers on the reversible film are yellow, magenta and red. Sections are mounted on the colour film and incubated.

Protein bound carboxyl groups

Carboxyl groups take red and blue colour. Red denotes scattered or few groups while the blue colour points to many closely spaced groups. The incubation of the section is performed in 0.1% 2-hydroxy-3-naphthoic acid hydrazide and stained with tetrazotised diorthoanisidine.

Protein bound sulphhydryl groups

The well known Barrnett and Seligman method was modified by substituting a monocoupler, fast blue RR for the dicoupler, tetrazotised diorthoanisidine. Sulphhydral proteins stained deep blue. These two methods compare vigorously with respect to detection and quantity.

Protein bound tyrosine

It shows greater red pin colour intensity when treated with a coupling reagent 1-amino-8-naphthol-4-sulphonic acid (S-acid) coupled with a 1% KOH solution. The process is known as protein diazotization followed by alkaline coupling.

Propionic-iron-alum haematoxylin stain

It gives excellent metaphase chromosome preparations. The nucleolus becomes bright red. In fungi, its ability to stain the centrioles bluish red enables one to follow their behaviour throughout preferably in squash preparations.

Prussian blue

Strong reducing groups such sulphydryls of the epidermis, ascorbic acid or the reduction of -S-S to -SH by ferric ferricyanide produce the precipitated pigment. Prussian blue. Thus prussian blue is ferric ferrocyanide.Blue precipitate with Fe^{3+} is liberated from ferritin and haemosiderin. Nuclei become pink or red. Iron can be removed before staining with aqueous oxalic acid.

Pyocyanin

It has a selective action. It is a synthetic dye and stains mucopolysaccharides blue.

Pyronin controls

Although complete understanding of pyronin reaction for RNA is unclear, the validity of the stain is established by using ribonuclease as a control. It indicates tissue basophilia. The stainable groups in tissue include the -PO_3 groups of nucleic acids and -COO- groups of the proteins. Since the dye-tissue interaction is electrostatic, the dye responds to the total charge density in the vicinity of stainable components. The bulk of total RNA in nucleolus, chromosomes and cytoplasm is protein bound and with suitable fixatives it is retained in the cell. Pyronin Y (G) is less diffuse and better differentiaated than pyrinin B Extraction of RNA in 5% aqueous trichloroacetic acid at 90-950°C for 10-15 min is a rewarding control reaction.

Pyronine Y-G and B

It is a specific and standard stain for RNA. It stains cytoplasm and nuclei pink to red. It is quite effective when used with methyl green which stains nuclei blue. This method for DNA or RNA is confirmed by specific removal of either of these substances by enzymatic hydrolysis or chemical extraction. the designation G emphasizes the gelblich shade of this red dye but this was later changed in America to Y for the yellowish shade. Pyronin B contains two ethyl groups instead of two methyl groups and produces a bluer shade.

Quanta corrected spectrometer

It measures action excitation and emission spectra in addition to fuctioning as spectrophotometer. Equal numbers of quanta at different wavelengths are adjusted electronically with control of the slit through a quantacorrected photocell or a light dependent resistor which is programmed by a cam connected to the wavelength drive (Per Halldal et al., 1975).

Quantitative histochemistry

It reduces the scale of biochemical methods in order that the data obtained from their determinations would reflect the morphological complexity of the tissue used. The aim is to reduce the amount of material necessary for a determination in order that differences between samples can be detected and perceived.

Quantitative video intensification microscopy (QVIM)

It measures fluorescence at the cell level. It uses a low light video camera which is interfaced to a microscope and to a microprocessor controlled a video digitizing system. One can limit the even point areas in the field for measu;rement. The technique responds linearly to light input and the results are reproducible to a high degree.

Quinalizarin

Magnesium will develop a bright blue colour after 12-24 hours. Both the reagents are to be fed to the shoot apices drop by drop. Quinalizarin with sodium acetate is one while the second is 10% NaOH solution.

Radioactivity

In the technique of autoradiography, radioactive compounds are exogenously supplied to plant or animal tissues and trace their eventual fate within the tissue. For example radioactive amino acids are used to localise or trace proteins, thymidine to study synthesis of DNA, and uridine to study RNA. These precursors are made readioactive by replacement of a normal atom by a radioactive atom such as P_{32}, S_{35}, C_{14} or H_3.

Ramazol brilliant blue

It stains nuclei light blue and red blood cells deep blue. The staining is selective and precise and it functions exactly alike to acid fuchsin, aniline blue and fast green for vegetable cells.

Red blue staining

DNA take red while RNA blue contrasting colour. Thus nucleic acids take double stain by basic-fuchsin-methylene blue; hence known as red blue double stain.

Rhodamine B

This is an amphoteric dye of the xanthene series. It fluoresces when excited by UV or green light. Lipids fluoresce white pink. In macerated tissues or by injecting into the xylem/phloem, the dye works specific.

RNA by methylene blue

A specific staining procedure imparts blue coloured reaction to RNA. The hydrated sections are stained in 0.02% w/v methylene blue in citrate buffer at pH 3.5 for 30-50 minutes. Controls include extraction of RNA by perchloric acid, trichloroacetic acid, or even hydrochloric acid.

RNA by pyronin alone

RNA stains bright red-pink by pyronin alone. Thus pyronin alone can be used to demonstrate RNA in fixed plant tissues.

RNA in situ

In days long gone by the only method to extract DNA leaving RNA *in situ* was the use of deoxyribonuclease. The chemical method implies a depurinization of DNA by mild hydrolysis with 10% formalin in saturated picric acid for 24 hours at 30°C. The apurinic is condensed with 10% aniline in 25% aqueous acetic acid for one hour at 25-30°C. RNA in nucleoli and cytoplasm is exclusively well preserved and is now open for only detection.

Rosindole reaction

Sections are first treatment with *p*-dimethylaminobenzaldehyde (DMAB) in acetic acid to which strong perchloric or hydrochloric acids are added. Prroteins containing tryptophane are stained brilliantly blue. Itt clearly demonstrates tryptophane moiety of proteins. The specificity and sensitivity on a sharp detection of the rosindole reaction are established.

Rubeanic acid

An insoluble black precipitate is produced on the tissues and cells if they contained copper. It is an excellent histochemical method to demonstrate the distribution of copper in plant indicators of underground water.

Ruthenium tetroxide

It is an extremely superior fixing agent for pollen mother cells (PMC) to detect chromonemata of the chromosomes. They appear dark brown against total gray cytoplasm in *Tradescantia, Limnophyton, Zosterra* PMC.

Rubber in guayule

Staining with oil red O in formic acid gave uniform deep red staining. Iodine-potassium iodide stains the rubber globules reddish yellow. Osmium tetroxide stains the rubber droplets black. Oil red and densyl chloride are used together for epifluorescence microscopic localisation of rubber.

Sakaguchi reaction

Sites of arginine rich basic proteins or histones stain orange-red. Absorption maximum of the end product is at 517 nm proves the validity of the reaction. High concentration occurs in nuclei but the stain starts fading after 12 hours.

Sambucyanin

Both deoxyribose and ribose nucleic acids stain bright red and dull green respectively when tissue sections are treated with 10% aqueous solution of sambucyan in counterstained by light green. Nucleic acids will be stained bright red to dull red with sambucyanin alone.

Santalin

The nuclei of animal and plant tissues are stained dark brown while the cell cytoplasm is pale brown. The stained slides do not fade for several yeas. This stain is extremely useful as a counterstain in in nuclear-cytoplasmic background counterstain with histochemical procedures.

Schiff's reagent

Decolourized para-Rosaniline (dye-fucshin) is its basic ingredient. The dye combines with an aldehyde to give red-violet colour. In acid solution with excess SO_2, fucshin is transformed into colourless leucofucshin N-sulphinic acid. Schiff was a German chemist who developed the reaction with aldehydes. Schiff's reagent consists of a solution of the colourless derivative of the dye in water. It is useful in histochemical studies as the final product is a methyl sulphonic derivative of basic fuchsin.

Schweitzer's reagent

It is the only control reaction for cellulose, as it dissolves quickly in the above reagent. The saturated solution is prepared by dissolving copper hydroxide in undiluted commercial ammonium hydroxide.

Sea water

Aqueous 0.1% neutral red mixed with sea water produces a conspicuous fluorescence of either lemon-yellow or blue-green in lipids or fats of plants and animals. The water-soluble, non-fluorescent dye predominating in acid solutions become deprotonised, lipophilic and fluorescent under neutral red and alkaline conditions.

Silver-methenamine reaction

It pinpoints aldehyde groups of carbohydrates by black silver deposits. The reagent is prepared by mixing freshly prepared 3% aqueous methenamine and 5% silver nitrate.

Silver nitrate-rubeanic acid

The above union stains calcium oxalate deposits black or brown when nuclei are stained green.

Silver staining of nucleoli.

In garlic meristem the nucleoli are stained with $AgNO_3$. It takes deep black colour. The technique helps to study tthe process of nuclear fusion. During interphase, nucleoli did not fuse but after the G1 phase of the cell cycle, there was immense nuclear fusion. If feulgen staining is performed first followed by $AgNO_3$ precipitation, the estimates of the absorption values will differ indicating fusion.

SITS

It is a specific cytochemical reagent for the outer components of the plasma membrane of the plant cell like leaf cells of Allium cepa or the intine of the pollen grain. It does not stain the cell wall. In proper concentration,

SITS serves as an indicator of cell vitality in tissue cultures. The SITS is 4-acetamide-4'-isothiocyanostilbene-2,2'-disulphonic acid.

Sodium bismuthate

It is a powerful oxidising agent. After oxidation with it if the sections are stained with leucofucshin, the sites of 1,2-glycol linkages stain lovely violet but yellow by nitrophenyl hydrazines. Like lead tetra-acetate if not only oxidises 1,2-glycols but also is able to reduce α-hyroxy acids to lower aldeydes.

Sodium tungstate

Certain phenols and tannins including gallic,ellagic and chebulagic acid, pyrocatechin, catechol pyrocatechol and pyrogallol stain light brown when treated with sodium tungstate dissolved in sodium acetate and water. When sections are transferred to aqueous $K_2Cr_2O_7$ the stain colour becomes prominent.

Solochrome dyes as metal indicators

These dyes include Eriochrome black T for calcium, Solochrome dark blue BS for aluminium and Solochrome azurine BS and S cyanin R for aluminium and beryllium are specific. Nuclei stain deep blue while basic and lipoproteins in various shades of red.

Spirit blue

The ubisch granules on the tapetal membrane stain selectively an intense bright blue and are beautifully revealed in an unstained back ground. In addition the pollen exine and sporopollenin also take a deep blue stain, by the sprit-soluble aniline blue dye.

Sporopollenin

It is a carotenoid materrial which forms the chief constituent of the pollen exine. It is highly resistant to extremes of decomposition. It ranks as the most resistant wall polymer produced by plants. Its formula is $C_{90}H_{129}O_{12}(OH)_{15.}$ Its deposition begins immediately after release of individual microspores out of the tetrad configuration. These grains can be clearly stained by saffranin dye or stained by nigrosin. It also stains strongly with osmic acid and reacts with PAS treatment. On account of its high refractivee index it appears dark under phase contrast microscope.

Stining mitochondria

The mitochrondria stand out sharply purple-red with peripheral wall green when stained with aqueous acid fuchsin and counter-stained with 1% aqueous light green.

Standardization

The presence of minor quantities of related dyes or other impurities change the specificity of the dye. The presence of unknown impurities in a dye will be a limiting factor in the understanding and quantitative interpretation of dye-substrate interaction. Performance tests on each batch of a product are therefore carried out in addition to optical and chemical tests.This needs for the standardization of stains. The commission ertified stains are marked "CC.". The formula for the calculation of the dye content of any sample is:
dye molecular formula

$$\text{percent dye} = \frac{\text{CC } TiCl_3 \text{ used x normality factor x molecular weight of dye x 100}}{\text{weight of sample x no. of hydrogen equivalents X 100}}$$

or

$$= \frac{\text{CC } TiCl_3 \text{ used x normality factor x 10 x 100}}{\text{CC of N/10 } TiCl_3 \text{ required per gram of dye x wt. of sample}}$$

Starch

On staining a preparation of potato tuber in potassium iodide-iodine (KI-I_2) solution, short chain starch molecules stain red-brown while long-chain starch molecules stain deep blue. When viewed with polarised light starch grains show distinctive birefringence. Starch in process of being formed stain different shades ranging from red to violet.

Steroids

A saturated solution of antimony trichloride ($SbCl_3$) in 60% $HClO_4$ produces a specific red colour of steroids-viz. diosgenin and yamogenin. The plant tissues very commonly used are from stems and roots of *Solanum Khasianum* L.

Sticky wax

It is an embedding medium for hand tissues containing a mixture of carnauba wax, gum dammar and natural rosin. Its infiltration is quick, and its melting point is about 20°C higher than paraffin. Lignified and calcified elements are processed well.

Suberin

It consists of aromatic phenols and long-chain aliphatic acids. Thus it shows lipid-like properties when sections of woody stems of *Anona* or *Thuja* are irradiated with UV-light, suberin fluoresces yellow.

Succinate dehydrogenase (SDH)

It is a primary anaerobic dehydrogenase, readily soluble iron flavoprotein. It is a Krebs cycle enzyme. It catalyzes the reversible oxidation of succinate to fomarate. SDH activates the hyudrogen atom of succinic acid and affects their transfer ultimately to tetrazolium salt.The enzyme converts the tetrazolium to the water insoluble formazan. The amount of formazan produced over short periods is proportional to enzyme activity. Mitochodria are intensely stained bluish magenta. Use of nitroneo tetrazolum chloride gives fine granular deposits of formazan by SDH.

Sucrose

An opaque precipitate with methanolic solution of barium hydroxide in regions containing sucrose offers a specific test. It is due to the formation of barium saccharate. It forms an insoluble complex with barium hydroxide in an alcoholic medium. It is a simple and rapid method to specify its distribution in the tissues of sugar beet.

Sudan III

Fats and fatty acids, which are neutral impart blood red colouration. Dissolve the stain in 50 to 70% alcohol lipids tend to dissolve in absolute alcohol.

Sudan IV

Lipids (neutral fats) stain orange to red. Staining is very critical. Counter staining with alum-haematoxylin imparts contrast by blue colour. The method is quit adequate for the detection of isolated fat cells.

Sudan black B

Lipids stain brown black or blue though rarely phospholipids appear brownish black. But extraction by ethanol, methanol-chloroform prior to staining will prove the specificity of the reaction on the mitotic and meiotic chromosomes and chromatin of interphase nuclei. Commercial samples contain many contaminants or deterioration products. Sudan black B diffuses into the interior of fat droplets and the advancing stain can be traced under a light microsocope.

Sudan VII B

It stains lipids good bluish red colour. The use of aqueous tween solution as the solvent for fat staining is quite new but interesting. It is also known as Sudan red 7B.

Sugar acetocarmine staining

An *in vitro* culture solution of 15% sucrose, 360 ppm calcium chloride dihydrate and 120 ppm boric acid forms a germinating medium. On part of fresh pollen is stirred with 9 parts of the solution and kept for 5-7 hours. One part is then mixed with two parts of acetocarmine stain. The sugar in the mixture maintains the colour contrast between the pollen cytoplasm (light pink) and the nuclei (reddish purple), decreases the bursting, increases the pollen diameter, and stabilises the pollen configuration. Monocotyledonus pollen responds well and contains bright 3-cells.

Sulphdryl groups

The orientation of -SH proteins into spindle strands is associated with transformation of intramolecular disulphide bonds into intermolecular bonds via their reduction to sulphide groups. It is a two step reaction. Both the reactions are reversible. Mercuic ions react specifically with proteins containing sulphhydryl groups.

Sulphydryl proteins

Sites of proteins containing sulphydryl groups stain sky blue by potassium ferricyanide freshly prepared and mixed with ferric sulphate. Lack of blue colour after blocking the groups in a saturated aqueous solution of $HgCl_2$ or saturated phenyl mercuric chloride support the reaction.

Synergids

These two cells flank the female gamete. They secrete either a chemotactant or certain enzyme which direct the pollen tube. Their detection confirms the production of simple water-soluble carbohydrates.

Tannins

It comprises a group of astringent substances dissolved in cell sap. It is localised in bark, unripe fruits, leaves and galls. Ferric acetate or ferric chloride imparts a blue colour. Catechol tannins and phenols develop a cherry-red colour by the nitroso reaction. They can be precipitated by basic lead acetate too.

Television flurescence microspectrophotometry

It permits the detection of even extremely minute quantities of substances in the living cell.

Tetra-azotised o-Dianisidine

It stains phenols, tryptophane and histidine pleasing red. It is dissolved in veronal solution by grinding the dye powder.

Tetrahydrofuran (THF)

It is extremely useful as a deparaffinizing and dehydrating agent in staining schedules. It is less toxic, rapid, preserves clarity of contents and maintains stain acceptance in plants. It is superior to tertiary-butyl alcohol. It is also useful as a component of fixatives.

Tetrazole reductase

Locations of the above enzyme reveal the deposits of insolube colour formazan, by the action of derivatives of tetrazole CH_2N_4.

Thallium

It is histochemically detected in the sections by the sulphide–selenium silver method. Small black granules designate the sites of thallium. The sections are treated with gaseous H_2S for 15 minutes and with 20% ammonium sulphide saturated with powdered selenium for 10 minutes.

The PA-TSC-SP test

Aldehyde groups are identified by silver deposits. It is a complex solution of aqueous 1% periodic acid plus thiosemicarbohydrazide 1% in 5% aqueous acetic acid and aqueous 1% silver proteinate at pH 4.4 or 9.2.

Thio compounds

Regions containing thio compounds appear pink by treating the sections of mustard, cabbage, onion and garlic by 2,3-dichloro-1,4-naphthoquinone in ethanolic ammonia. The allylisothiocyanate in mustard, 4-methylthiobutylisothiocyanate in cabbage, allypropydisulphide in onion and diallyldisulphide in garlic give a graded pink colour product in presence of ammonia.

Thionin orange-G

It is a specific stain for fungal parasites in plant tissues. The fungus mycelium and spores are purple against the yellowish green walls of the host. It locates *Plasmodiophora* in cabbage, *Perenospora* in pigwood and *Puccinia* on *Triticum*. The hyphase and spore walls are purple, middle lamella red, lignin green, cytoplasm pink, cellulose and cut in yellow.

Titan yellow

Thiazole quinalizarin yellow test is quite specific for determination of magnesium. The tissue produces outgrowths by the precipitation of a membrane that forms around the dissolved particles of phytates (organic phosphates) after $FeCl_3$ treatment. Magnesium develops a brick red colour after several hours.

Totuidine blue O

It is a blue cationic dye freely soluble in water but sparingly in alcohol. Nucleic acids stain orthochromatically at the appropriate pH. Tannins stain green-bright blue. Polyphosphates and sulphated compounds stain metachromatically even down to pH 1.0. The presence of phenolic substances is indicated by a greenish blue to green colouration which persists at low pH. With naphthol yellow S, it results in striking spectrum of colours including various shades of blue, yellow, green, brown and red.

Total lipids

All the Sudan stains are specific for lipids. The fats, oils and waxes stained by Sudan III and IV blue to black. Sudan black B stains the free fatty acids and phospholipids dull black.

Transmission

This is measured by the use of a cytophotometerr with a monochromatic filter. The amount of light that comes through when the dye-metabolite complex is present can be expressed as a percentage of the amount that comes through when it is not. This percentage is known as transmission at the particular wavelength used. For instance instead of saying that the transmission at a particular wavelength is 20/100 of the incident light. We can turn the fraction upside down and use the number 5 to represent the absorption.

Trichloroacetic acid (TCA)

It removes nucleic acids from the chromosomes. It finds out the precise seats of DNA and RNA followed by methyl green pyronin. The amount of DNA present is calculated by the deoxyribose sugar. Total nucleic acid content-minus the DNA content gives RNA present. DNA will stain blue and RNA red. Blue nucleus has a red nucleoulus and cytoplasm.

Triethylene glycol

It softens lignified wall of the wood and hence permits cutting sections without tearing or crumbling. This treatment is flexible and adaptable to a wide variety of tree species. This is the only chemical that will readily dissolve lignin if catalytic amount of a mineral acid is added to it.

Triple stains

Nuclei stain blue to green, polysaccharides red and proteins yellow. Plant material sections are stained successively in Feulgen reaction (Schiff's reagent or Azure A); peeriodic acid-Schiff (with basic fuchsin-Schiff reagent) and 0.02% solution of naphthol yellow S in 1% acetic acid. The staining reactions are specific and excellent for coloured photography.

Trypan blue

Nucleus stains red with this dye in plant cells. But the red fraction of trypan blue is highly selective for degenerating or pycnotic nuclei or nuclei undergoing quick and rapid changes. The red fraction of the dye is also selective for basic proteins. Evans blue is more specific than Trypan blue to determine the percentage of viable cells. Intact live cells prevent the trypan blue dye from entering and so remain unstained. In aleurone grains crystalloid becomes strongly red violet and globoid totally unstained.

Tryptophane

The reaction consists of treatment with nitrous acid followed by N-(1-naphthyl) ethylene-diamine dihydrochloride. A purple product appears in the presence of proteins containing trptophane. Controls contain treatment with 7- disulphonic acid or 1,6 dihydroxynaphthalene coloured product is not indicative of tryptophane.

Ubisch spheroids

The tapetal cells contain granular bodies often attached to the membrane bound organelles. These bodies disappear from the inside of the cell and thereafter appear in the plasma membrane. From here they are located first to the inner tangential walls and then to the exine of the microspore. In *Tradescantia* it is of microspore origin only. This aspect requires further attention.

These bodies are stained by saffranin-fast green. Are these polymerrised into sporopollenin? Spirit soluble aniline blue stains these grains.

Ultra-histochemistry

In histochemistry using light microscope the chief aim is to produce a visible coloured end product, well localised at the site of the substance in a cell therefore insoluble. In ultra-histochemistry the purpose is to produce a well localised end product which is detectable by an electron microscope; therefore the most important property of the reaction product is that is must be electron dense.

Ultra-violet Cytophotometry

It is one wavelength and two area method. The operating wavelengths are isolated by means of MM 12 double monochromator equipped with quartz prisms. The spectral transmittance in the 250-315 nm range is 15-20 per cent and the bandwidth at 265 nm, with a fully open slit is 3 nm. the exit slit of the monochromator is focussed on the microscope mirror by a system of quartz lenses fitted with small field work. The stray light is reduced by a couple of circular holes. The accuracy of the procedure equals 93 per cent. There is no correlation between cells size and DNA conent (Garcia, 1973).

Units of measure in histochemistry

Weights

1 μg = 10^{-3} mg = 10^{-6} g

1 mμg = 10^{-6} mg = 10^{-9} g

1 μμg = 10^{-9} mg = 10^{-12} g

Volumes

1 μL (mm^3) = 10^{-3} mL = 10^{-6} L

μL = microlitre

μg = microgram

The measurements on the cytophotometer of fluorimeterr are comparative and not absolute.

Units 1 mole

1 pickomole = 10^{-12} mole

(p mole)
1 nanomole = 10^{-9} mole
(n mole)
1 litre
1 manolitre = 10^{-12} litre
Gram umits
1 nanogram = 10^{-9} gram
1 pickogram = 10^{-12} gram
Femto = 10^{-15}
Atto -10^{-18}

Unna's stain

The unique combination of methyl green pyronin is famous as Unna's stain. Cells treated with Unna's stains have red nucleoli and cytoplasm but nuclei blue-a striking contrast. There are two procedures one by Brachest (1953) and the second by Taft (1951). Results of both staining are the same.

Unsaturated lipids

The unsaturated lipids stain black with 1% solution of osmium tetroxide in distilled water.

Vanillin test

A red colour is produced when aldehyde groups in the vanillin condense with phenols in the tissue.

Viscol

It is a permanent mounting medium, a limpid fluid and somewhat viscous. It forms a liquid soluble in water which after a few days hardens under the cover glass. It is a phenyl derivative of viscose in a solution of glycerol. It hardens through the loss of water and can be softened again by adding a drop of water.

Vital dyes

These offer a technique or a means of simple observations on living cells. The use of Rhodamine 13 and 6G for transport movement in conducting tissues, viable cells prevent Evan's blue from entering the cell, cresyl blue, neutral violet and neutral red stain reveal viability of cells and tissues, methyl violet, dahlia and mauvine for movement in staminal hairs of

Tradescantia, and triphenyltetrazolium chloride for seed life and germinability are employed on plant material.

Vital staining

It is a means of a number of simple observations on living cells or parts of them. Neutral red for vacuoles, Janus green for mitochondria, methylene blue for axons are common. It affords a supravital medium in determining morphogenetic movements in zygote, embryo sac etc.

Vitamin C

Blackened grains or red brown colloidal silver occur at the sites of ascorbic acid. It will not demonstrate free ascorbic acid as it diffuses quickly from the sections. Sections are exposed to H_2S to reduce dehydroascorbic acid to ascorbic acid, followed by treatment with 10% silver nitrate solution in 3% acetic acid.

Xanthoproteic reaction

Yellow colour turns red in the presence of ammonia to root and shoot apices sections warmed on a slide with a drop of concentrated nitric acid.

X-bodies

Darkly stained bodies detected in the pollen tube discharge in the synergid. Its chemical nature is yet unknown. The tube cytoplasm contains numerous small PAS- positive spheres (van Went, 1970). In cotton they also stain for DNA.

Xylem

It is the wood or vascular tissue that provides mechanical support and conducts water and mineral salts. DNA-Feulgen content in *Arisaema* (snake lily) metaxylem showed that nuclei go through a series of DNA doublings correlated with nuclear volume doubling and cell volume increase at least up to 16 or 32 ploid level. The relative DNA content was determined by Schiff's reagent measured by two wavelength method.

Xylenol orange

It is a fluorochrome useful in polychrome sequential labelling of calcifying tissues. There is a clearcut contrast to that of known fluorescent dyes. The fluorescence is located at the same site as that of tetracyclines, fluoresceins and calcein blue.

Yeast fixation

The fixative for yeast cells dimethyl sulphoxide-acrolein-glutaraldeyde preserves all cytoplasmic features, lipid deposits, ribosomes etc. It is the best permeating agent allowing maximum penetration of the cell wall without disrupting cellular fine structures and organelles.

Zinc-Chlor-iodide

It is used for cellulose detection. An acid or a strong ionic solution disrupts the hydrogen bonds, maintains the molecular configuration and iodine accumulates in the enlarged space imparting blue colour. However both the tests (IKI-H_2SO_4) remain approximate.